Human Memory and Material Memory

Christian Lexcellent

Human Memory and Material Memory

 Springer

Christian Lexcellent
Département de Mécanique Appliquée
FEMTO-ST Institute
Besançon, France

ISBN 978-3-030-07608-5   ISBN 978-3-319-99543-4   (eBook)
https://doi.org/10.1007/978-3-319-99543-4

Based on a translation from the French language edition: Mémoire humaine et Mémoire des matériaux by
Christian Lexcellent Copyright © Cepadues 2018 All Rights Reserved.
© Springer Nature Switzerland AG 2019
Softcover re-print of the Hardcover 1st edition 2019

This Springer imprint is published by the registered company Springer Nature Switzerland AG
The registered company address is: Gewerbestrasse 11, 6330 Cham, Switzerland

Preface

The present book attempts to establish whether a link between human memory and material memory such as the memory of shape memory alloys is relevant.

It challenges the strength of the boundary between these two concepts of memory and wonders whether "memory traces" in the brain may be related to the defects generated in metal alloys when they are trained.

The book first provides a brief historical record of works devoted to human memory, from the ancient Greeks to the present day.

Then, it turns to neuroimaging. With its development, neuroscientists believe that they can explain the whole brain behavior, including the memory process.

Then, following Paul Ricoeur's work, the philosopher's point of view is explored, which adds sensitivity, thought, and humanity to the technical brain. In conclusion, the link connection between the two types of memory is presently still difficult to prove and establish, but the reader is better equipped to understand and continue this open debate about human memory.

The link between the two types of memory is presently difficult to establish.

Besançon, France   Christian Lexcellent

Acknowledgements

The author would like to thank Mr. Dominique Banet, priest and philosopher, who encouraged him to develop the idea of a continuum between human memory and memory of materials.

Contents

Chapter 1
Introduction: A Few Questions

Abstract A few questions about memory' creation.

1. How can we create memory, whether for humans or for a material?
How is memory "acquired" ?
A partial answer can be that it is to make a person or a metal alloy "undergo" or "live" a STORY.
For men, memory can be collective (the "butchery" of the First World War, the Holocaust, the soccer world cup in 1998 in France ...) or individual.
Memory is inherent to any human species (and animals?).

2. Two parallels can be drawn between "Memory, History and Forgetfulness" (see Paul Ricoeur's book Ricoeur (2000)) + forgiveness!, on the one hand, and "material, training, and amnesia," on the other hand.

3. Memory is "I remember" Do "men deprived of memory" exist?
Apparently not.
"Because a man without memory is a man without life, a people without memory is a people without a future." (Marshal Ferdinand Foch (1851–1929)).
"A man without a past" is a film by Aki Kaurismaki (2002).
The story runs as follows: a man arrives in Helsinki, where he is attacked by a gang and becomes amnesic.
Then, he rebuilds his life with the help of homeless people who live in the city and the Salvation Army.
Should we remember our personal, collective story? Not all our recollections are fake, far from it!
"The Soft Watches", by Salvador Dali
"The Persistence of Memory" is a 1931 surrealist painting by Salvador Dal. It is an oil on canvas generally known as "The Soft Watches" by the general public, and one of the painter's most famous works.

C. Lexcellent, *Human Memory and Material Memory*,
https://doi.org/10.1007/978-3-319-99543-4_1

"Although I do not understand the meaning of my paintings while I paint them, this does not entail that they have no meaning."

Reference

P. Ricoeur, *La mémoire, l'histoire, l'oubli*. Editions du Seuil, Points Seuil (2000)

Chapter 2
An Attempt to Define Memory

Abstract The memory can be considered as individual, as a modal model, as regards neuroscience, psychoanalysis, and for materials.

If, first of all, we want to venture toward a technical approach to individual memory, we can define it as follows: "Memory is a biological and psychic activity that makes it possible to store, preserve, and retrieve information." Synonyms are "reminiscence" or "recollection."

This definition is often used in cognitive psychology:
The classical cognitivist current usually includes in the term memory the processes related to encoding, storing, and retrieving mental representations.
In ancient Greece, memory was of such great importance that Mnemosyne, goddess of memory, was the mother of nine muses, goddesses who presided over knowledge: Clio was beneficial to history; Euterpe to music; Melpomeni to tragedy; Erato to poetry; Ourania to sciences; and so on. There was a cult of Mnemosyne, and people who wanted to improve their memory went on thermal cures deemed to be beneficial.

As regards the modal model of memory:
The most influential structural model of memory is the modal model, which divides memory into three subsystems: a sensory record, short-term memory, and long-term memory. In the memory of actions and motor coding, the action benefits from a triple coding: verbal, imaged, and motor.

As regards neuroscience:
Neural plasticity is fundamental to explain the various forms of memory. We can note that the term "plasticity" was wisely/appropriately borrowed from the mechanics of materials. The Hebb rule, also called theory of neural networks, was established by Donald Hebb in 1949 (Hebb 1949): it is used as one the most important mechanisms underlying memorization. Zones of convergence–divergence are likely places of recording and reproduction of recollections and of other forms of long-term memory.

As regards psychoanalysis:
In psychoanalysis, the term memory is connected to metapsychology and in the first place to the first topic and thereby to the notion of the unconscious,

© Springer Nature Switzerland AG 2019
C. Lexcellent, *Human Memory and Material Memory*,
https://doi.org/10.1007/978-3-319-99543-4_2

of repression. Metapsychology is the set of theoretical concepts formulated by Freudian psychoanalysis since the "Project for a Scientific Psychology" (written in 1895 and published in 1948) (Freud 1948), "The Interpretation of Dreams" in 1900 (Freud 1900), and "Instincts and their Vicissitudes" , "Repression" and "The Unconscious" are three texts dating back to 1915, until "Some Elementary Lessons in Psycho-Analysis" in 1940 (Freud 1940).

As regards materials:
Some materials acquire memory after a training process. They are called "shape memory alloys" because they remember the geometric shape that they were given by a suitable thermomechanical treatment. Those are nickel–titanium or copper-based ternaries of the Cu–Al–Zn genus ... (Lexcellent 2013).

Benveniste's memory of water was rather the subject of a controversy that we will address later.

References

S. Freud, *Abrégé de psychanalyse*. (PUF (2010), 1940)
S. Freud, *Esquisse d'une psychologie scientifique* (1948)
S. Freud, *L'interprétation des rêves (1990), PUF, 2005*. (PUF (2005), 1900)
D.O. Hebb, *The Organization of Behavior*. Wiley (1949)
C. Lexcellent, *Les alliages à mémoire de forme*. (Hermes-Lavoisier, 2013)

Chapter 3
How Has Memory Been Visited Over Time?

Abstract Fundamental character for ancient Greeks. In Rome, the art of memory was developed for utilitarian purposes. Memory is as basis of knowledge and consists of images. The Middle Ages revealed themselves in the Trivium and Quadrivium. Latent distinction between impressions (sensation) and ideas.

Mnemé (memory) mnema (monument to remember) mnemeion (reminiscence) lethomai (I forget), the diversity of terms relating to memory attests the fundamental character it had for ancient Greeks (Lieury 2013, p. 10).

The invention of the first memory technique is attributed to Simonides of Ceos, a poet of the fifth century B.C. (−556, −467). Legend has it that Simonides was the only survivor of the collapse of a banquet hall and had to remember where the guests were seated to identify them. He deduced the famous "method of loci," which consists in memorizing the objects in the form of images and arranging them mentally in places (Fig. 3.1).

According to Plato (−427, −347), memory is an innate knowledge and not a conservation of lived experiences. Any evocation is a reminiscence of the soul.

By stating that "memory is from the past," Aristotle drew attention to the temporal aspect of memory phenomena. It is not just that memory images relate to people and things and places in the past, but these memory images have a before and an after in their order. Aristotle added to the discussion of the distinction between mneme and anamnesis the distinction between recollections that arise unexpectedly and recollections that require an effort to bring back. Unfortunately, neither Plato nor Aristotle was able to solve the aporia of memory (Reagan 2008).

As the author of the first treatise on memory (Aristote 1866), Aristotle considered that the heart was the seat of intelligence. This definition is dear to cognitive psychology:

The classical cognitivist current usually includes in the term memory the processes of encoding, storing, and recovering mental representations of courage and memory, and hence the expression "know by heart" which seems to have been used for the first time by Rabelais in the sixteenth century.

In Rome, the art of memory was developed for utilitarian purposes, especially to plead.

Fig. 3.1 Using the zodiac for the method of loci

Three treaties survived the destruction of the Roman Empire by the barbarians: (i) "Rhetoric for Herennius", by an unknown author (to remember images, we must single them out, and give them "exceptional beauty or utter ugliness by representing ourselves some of them bloody, covered with mud, or covered with vermilion" (Lieury 2013, p. 26),"Orator " by Cicero (Cicéron 1869) and "Institutes of Oratory" by Quintilian.

In his book "Memory and Reminiscence" (Aristote 1866), Aristotle (-384, -322) developed remarkable ideas for his time:

- empiricism,
- the role of images,
- associations of ideas,
- associative mechanisms,
- recovery processes,
- the reference to the past, and

"Memory arises from the past", "one must say to oneself, inside one's own mind, that one has previously heard or perceived or conceived such a thing."

These three works contain the same original principles:

- memory is formed of imprints in wax (like certain defects developed in the microstructure of certain metals)
- as sight is the strongest of our senses, visual images are memorized best
- images have to be ranked for us to recall them (as in the "method of loci" dear to Simonides of Ceos).

Cicero's profession and fame (he was born in -106 and died in -43 BC) (Cicéron 1869) led him to be in touch with many people that he took as proof of the effectiveness of the method of loci in his work "Orator". According to the way characters were etched in wax, they also engraved images in selected locations to add what they wanted to remember.

- Memory is at the basis of knowledge and consists of images. Recollections are connected in regular files; the same concept was taken up in the current neo-conceptualism of artificial intelligence in which our knowledge is connected inside associative networks.
- Abstraction in memory: St Augustine (354–430) went beyond Aristotle's conception by showing that the too fragile images ("residuals of perception") do not build up memory. St Augustine developed the notion of conceptual memory. He was a great precursor of cognitive psychology. According to him, most mental representations are conceptual; we do not remember the event, but its verbal re-coding.
- The multiplicity of memory: for us to be aware of forgetfulness, our memory has to operate. The solution is that there is not one memory, as the philosophers once believed, but several. Memory is modular. St Augustine grasped this complexity, but he believed that memory was a deep den that led to god ...

Contrary to what Alain Lieury wrote (Lieury 1992) about the Middle Ages: "1,000 years of obscurantism associated with the Middle Ages, from the destruction of Rome in 410 until about 1416, when Quintilian's book was found. The first Bible was printed by Gutenberg in 1456, Quintilian's text was published in 1470," the Middle Ages revealed themselves in the Trivium and Quadrivium.

If we think of the Middle Ages and education, traditionally speaking there were seven liberal arts. Three of them—grammar, dialectic, and rhetoric—formed the Trivium. The other four—arithmetic, geometry, astronomy, and music—formed the Quadrivium.

Other people considered that the Trivium represented the three arts, and the Quadrivium the four sciences. The limitation to seven arts, their division into three arts on the one hand and four on the other hand, can be found in the works of Martianus Capella, Cassiodorus, Boethius, and their successors, where they answered the mystical preoccupations which were then mingled with conjectures on numbers.

Bede the Venerable, Alcuin, John Scotus Eriugena, Gerbert of Aurillac, and Fulbert of Chartres taught the seven arts or considered them in the succession indicated by the Trivium and the Quadrivium. But this was confined to the intellectual activity of men in the Middle Ages. In addition to theology and sacred books, to which all gave a large place, they studied history, physics, philosophy, metaphysics (or morality), medicine, and later law (canon or Roman), alchemy, etc. The Trivium and Quadrivium were only part of medieval education.

The Trivium was a systematic method of critical thinking used to determine facts and certainties among the information perceived by the five senses: sight, hearing, taste, touch, and smell. For the medieval university, the Trivium was the lower division among the seven liberal arts, and gathered three of them: grammar, dialectic, and rhetoric (input, process, and output).

The etymology of the Latin word Trivium is "the place where the three paths meet" (tri + via); therefore, trivium subjects were the foundation of the Quadrivium, the upper division in liberal arts education in the Middle Ages, including arithmetic (pure numbers), geometry (numbers in space), music (numbers in time), and astronomy (numbers in space and time).

Around the 1500s, towns and trade developed. Memory was treated by some people as a mystical mission to find the magic keys of memory, and in a utilitarian way by others ... A cultural mixture occurred through highly diverse currents, namely, magic systems, neoplatonism, the hermetic philosophy of Hermes, and the cabal or secret Jewish science.

Mystics (Ramon Llull, Giulio Camillo, and Giordano Bruno) had the illusion of finding the keys to divine memory:

Ramon Llull used concentric wheel systems with numbers and letters and symbols that inspired coding processes.

An example of ball joints used to encrypt secret messages is given in Fig. 3.2.

Ball joints were used by the mystic Giordano Bruno in a frighteningly complex way in his work "The shadows of ideas" in Paris in 1582 (Bruno 1583). Figure 3.2 displays a circle formed of four ball joints, a system inspired from Llull or Trithemius. Each ball joint (or wheel) had two entries: the first was alphabetic and composed of 30 boxes corresponding to 30 letters (the Latin alphabet + Greek and Hebrew letters). Each of these 30 boxes was subdivided into five parts corresponding to five vowels.

Giulio Camillo used the "method of loci." The locus was the amphitheater built around the magic number 7, which has a symbolic importance in the field of memory.

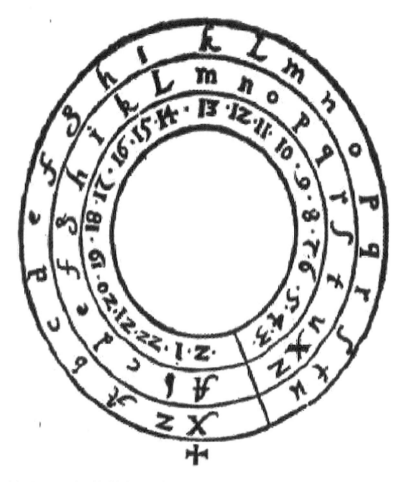

Fig. 3.2 An example of ball joints used to encrypt secret messages

The representations of Giordano Bruno's memory appear in two books "The Shadows of Ideas" (Bruno 1582) and "The Seal of Seals" (Bruno 1583) (an allusion to the imprints of memory, like seals in wax). Among the 30 seals, seal n° 1 was made of memory and imagination, whose vast folds could provide room for images. Seal n° 8 was "the sky" to be able to engrave the order and series of images from the sky.

This "knowledge memory" concept inspired many mystical authors of the Renaissance who sought the mystical "key" to access divine knowledge. Because he pursued into this quest, Giordano Bruno perished on the stake of Roman inquisition (Lieury 2013, p. 20).

In strong opposition to the tradition of Lambert Schenkel's methods of artificial memory, Descartes's "Discourse on the Method" (1596–1650) was based on categorization, which consists in putting items that resemble one another in a same group.

The mathematician Descartes celebrated reason and condemned the methods involving images and magical wheels to remain crammed in their sulfur-smelling lairs. Thus, memory was remarkably absent from Descartes' works. He mistook Schenkel's absurd methods with memory as a whole (Lieury 2013, p. 64).

As far as mathematics is concerned, we can cite Richard Grey as the inventor of the letter-digit code (Grey 1730). It was first taught in the mathematics class of Pierre Hérigone, who proposed to replace numbers by letters (consonants or vowels) and syllables (Hérigone 1644).

Leibniz, who invented among others infinitesimal calculus (differentials and integrals), was interested in the letter-digit code.

The "empiricist-associationist" current included philosophers whose ideas greatly differed.

For Hobbes (1638–1679), lived experience produced memory, and images were "associated" as in our associations of ideas.

Locke (1632–1704) introduced "the association of ideas."

"Let us suppose that the mind were like a sheet of blank paper, void of all characters, without any ideas: how would they come up? To this, I answer in one word: through experience."

Berkeley (1685–1753) postulated that psychological functions were not products of experience but mediated by clues.

He was therefore the first to introduce the idea of abstraction based on the clues of the situation.

Hume (1711–1776) made a latent distinction between impressions (sensations) and ideas that were weakened impressions to him. He also stated the need for association.

From a dogmatic viewpoint, the culminating point of associationism was reached by James Mill (1773–1836): for him, everything was association, thought followed from thought, ideas followed from ideas, incessantly ... "He was a reductionist since he reduced all the laws of association to one: contiguity."

John Stuart Mill (1806–1873)—James's son—added "similarity" to the laws of association.

In chemistry, the properties of a given compound, e.g., water, cannot be deduced from the properties of the atoms that make up the compound, oxygen on the one hand and hydrogen on the other hand in the present example. In these cases, we have to introduce "interaction."

The phenomenon that we call "interaction" suggested John Stuart Mill the theory of "mental chemistry" according to which complex ideas are not reduced to simple ideas. This theory very explicitly announces the theories of the Gestaltists, for whom a complex structure, for example, a word ("water" for example), is not reduced to just the sum of the parts. Therefore, it is the first theory of organization.

Let us illustrate this with images of the water molecule, the hydrogen bond, and ice.

As shown in Fig. 3.3, the water molecule is very simple; it consists of two hydrogen atoms and one oxygen atom.

Fig. 3.3 A water molecule

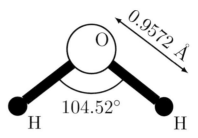

Fig. 3.4 Water in the
liquid state

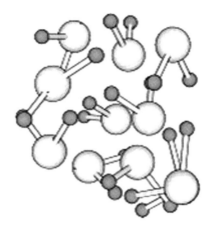

The connection between two water molecules shown in Fig. 3.4 is more complex, not to mention the hexagonal structure of ice (Fig. 3.5) obtained by simply cooling water.

Where can memory fit into these crystallographic structures?

We often talk about memory through its deficiencies. In his book "Les Maladies de la Mémoire", Theodule Ribot (1901, 1st edition, 1881) (Ribot 1901) developed a theory of aging that went against Darwin's theory of evolution (1859) (Darwin 1859). Korsakoff's syndrome due to chronic alcoholism (and causing lesions of the hippocampus) and Alzheimer's disease is mentioned.

The German scientist Hermann Steinhaus was the first man to have ever measured memory. In his historical book, "The Memory", a contribution to experimental psychology, he developed different techniques for learning and relearning. His most famous experiment remains the first quantitative demonstration of forgetfulness (Lieury 2013, p. 98).

The Russian physiologist Yvan Pavlov discovered conditioning (thanks to studies on dogs' digestion). Conditioning is of such importance because it is the mode of learning of the autonomous nervous system, a part of our nervous system (Lieury 2013, p. 104).

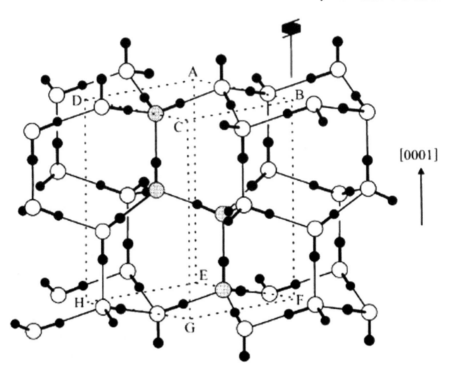

Fig. 3.5 Crystallographic structure of ice: oxygen atoms are represented by white or gray circles, hydrogen atoms by black dots. The structure is globally hexagonal

References

Aristote, *De la mémoire et de la réminiscence* (Ladrange, Traduction de Jules Barthélemy-Saint-Hilaire, -55 avant JC, Paris, 1866)

G. Bruno, *Ars memoriae dans de Umbris Idearum* (Bibliothèque Nationale, 1582)

G. Bruno, *Les ombres: Les sceaux dans Opera latine conscripta* (Bibliothèque Nationale, 1583)

Cicéron, *De oratore, L'orateur traduction* (M. Nisard, 55 avant JC, 1869)

C. Darwin, *On the Origin of Species by Means of Natural Selection, or the Preservation of Favoured Races in the Struggle for Life* (John Murray Londres, 1859)

R. Grey, *Memoria Technica: Or A New Method of Artificial Memory 1re édition 1730* (Cambridge University Library, 1812)

P. Hérigone, *Cours mathématique Tome 2* (Bibliothèque de la Sorbonne Paris, 1644)

A. Lieury, *La mémoire* (Mardaga, 1992)

A. Lieury, *Le Livre de la Mémoire* (Dunod, 2013)

C. Reagan, Réflexions sur l'ouvrage de Paul Ricoeur: la mémoire, l'histoire, l'oubli. Transversalités **106**, 165–176 (2008)

T. Ribot, *Les Maladies de la Mémoire* (Félix Alcan Paris, 1901)

Chapter 4
How Does Memory Work?

Abstract Practice of lobotomy until 1991. For the memory investigation, "The neuroscience" disciplinary concept is introduced and the neuroimaging as a fundamental tool. Types of memory are explained. Memory works mainly through three processes: encoding, storage, and recovery. Nature memory is located in the brain.

4.1 The Rise of Cognitive Neuroscience

Foreword (Gaussel and Reverdy 2013):

In 1949, the elite of Medicine honored the Portuguese physician Egas Moniz, Nobel Prize for his work on "pre-frontal leukotomy applied to the treatment of certain psychoses and mental disorders," later renamed lobotomy. How did it happen that this sinister removal of a part of the brain passed through the filters of the Academy? In 1935, based on observations on monkeys, this doctor proposed a novel treatment for certain mental pathologies by partially disconnecting the pre-frontal lobes from the rest of the brain. On November 11, 1935, the first patient, a 63-year-old prostitute suffering from melancholy and paranoia, was operated on. Nineteen more patients followed; out of the 20 people treated, the doctor announced 7 "cured," 7 "improved," and 6 "unchanged" patients. Although the sample was very small, it was enough to launch a practice.

In most Western countries, lobotomy became a common practice: thousands of patients were operated on despite the protests of many psychiatrists. In the United States, Walter Freeman "perfected" the method: instead of boring holes on both sides of the skull, he passed through the eye socket. In 1952, the discovery of the first neuroleptic drug changed the deal: surgery gave way to chemistry, at least for most patients, while Moniz's heirs still struck here and there. In some countries, such as China, brain surgery remained commonplace. In France, the last "official" lobotomy dates back to 1991 (Herzberg 2017).

And which parts of the human brain play a key role in memory? (LESTA-cognitive health study lab of seniors 16-02-2014)

© Springer Nature Switzerland AG 2019
C. Lexcellent, *Human Memory and Material Memory*,
https://doi.org/10.1007/978-3-319-99543-4_4

In the last 50 years, neuroscience has developed considerably, thanks to technical progress allowing for very accurate indirect measurements of the activity of our brain (or that of animals). In parallel, cognitive sciences, with the theory of information processing (a model considering the human mind as a complex system that processed perceived information), have become the discipline that addresses the functioning of human or artificial knowledge, but "still have fuzzy outlines" (Chamak 2011).

4.2   Neuroscience or Cognitive Science?

The history of neuroscience is based primarily on the history of neurons. Following the introduction of the concept of the neuron as a nerve cell by Waldeyer in 1891, a branch of experimental physiology, namely, neurophysiology, was created in England in the 1930s. A first wrangle (others followed) opposed the proponents of neurological physiology to those of neuronal physiology until neurobiology imposed itself in the 1950s (by focusing research on the cellular level instead of nervous circuits). This led to the birth of neuroscience in 1969, with the creation of the American neuroscience society (its French counterpart was created in 1988).

The "neuroscience" disciplinary field concerns the study of the functioning of the nervous system from the most elementary aspects (molecular, cellular, and synaptic) to the most integrative ones that deal with the behavioral and cognitive functions (see the French CNRS website). Neuroscience historically started with the study of mental functions affected by pathologies. It was based on Gall's hypothesis, which postulated at the beginning of the nineteenth century that the brain was separated into different organs, and each controlled one of the faculties of the mind. This idea corresponded to cerebral localization (a given mental function corresponded to one area of the brain). It is still very present in research today, even if scientists are currently investigating the hypothesis of a mental function that comes from cooperation of several neurons organized into networks not always located at the same place in the brain (Tiberghiem 2002).

4.2.1   A Little Bit of Neurology to Better Understand It All

Cognitive or cognition (knowledge) sciences form a field at the interface of neuroscience, psychology, linguistics, and artificial intelligence (some add anthropology, philosophy, or epistemology, see Valera (1997)). Therefore, this field belongs to both the natural and human sciences (Tiberghiem 2002). The notion of cognition is at the root of cognitive science. It dates back to the mid-twentieth century and refers to the biological function producing and using knowledge. It includes modes of operation (called mechanisms, procedures, processes, and algorithms) of different kinds that are processed by specialized and controlled modules. From the point of view of cognitive science, neuroscience is one field among others in the study of cognition, just like cognitive psychology. Cognitive neuroscience can therefore be defined as a "set of disciplines whose purpose is to establish the nature of the relationship between cognition and the brain" (Tiberghiem 2002). To go further, Chamak (2011) drew a historical summary of the birth of cognitive sciences and discussed the different disciplines involved (see also Dworczak 2004; Ecalle and Magnan 2005).

4.2.2 Neuroimaging as the Fundamental Tool of Cognitive Neuroscience

The advances in neuroimaging of the last few years give us the illusion of seeing brain activity in real time, which paves the way for many speculations. A method widely used in neuroimaging is the subtractive method: two different tasks performed by a same person, involving or not a given mental process, are compared, and the brain areas that show differences in activation between these two tasks are recorded. The hypothesis is then to consider that the activated brain areas are involved in the mental process under test. In cognitive neurosciences, only noninvasive imaging methods are used. These are indirect methods involving measurements of the electrical or magnetic activity of the brain by electroencephalography (EEG) and more particularly the abovementioned potentials (ERP for event-related potential), and magnetoencephalography (MEG), or measurements of blood flow by functional nuclear magnetic resonance imaging (fMRI) and positron emission tomography (PET scan).

Medical imaging rather shows the defects of the brain, which does not have a corpus callosum, i.e., "cables" that connect the two brain hemispheres and make them communicate; the two hemispheres function as two independent "hard drives."

All these techniques require heavy computational treatment to "reconstruct" an image from these indirect measurements, and therefore skills and special training for the interpretation of the images thus obtained (Classes and Vialatte 2012). Experimental conditions, especially in fMRI and MEG, are often drastic, and experiments are expensive. The combination of the different techniques, which are often complementary, provides more and more convincing results of a same experiment. Some classical neuroscience techniques are also used to confirm results or to start experiments: neurosurgery, studies on brain lesions that prevent cognitive activity (thanks to transcranial magnetic stimulation), autopsies, invasive, or noninvasive studies on animals.

Although memory is a well-known function, it is quite complex. To better understand what it is, it is important to differentiate the different types of memory.

4.3 Different Types of Memory

- Lexical memory:

The information sent by the eye to the brain is not the image of an object but a barcode. This memory is thus a kind of library containing the files of all the words (Lieury 2013, p. 186). The memory of images is more efficient than the memory of words, which shows that the poet Simonides was right against the scholar Descartes who favored words over images! The strength of images on our brains is shown by the fact that it takes long months to learn to read whereas watching television is spontaneous (Lieury 2013, p. 180–182).

- Sensory memory:

Sensory memory is the memory that perceives external information. It maintains all the information perceived by our senses for a few milliseconds before it accesses other types of memory if the information is relevant.

- Short-term memory:

("machine–man" concept) Short-term memory holds information temporarily. It is an essential transition toward long-term memory. But not all the information contained in short-term memory will be kept in long-term memory. Indeed, a great part of it will fade away or deteriorate after a few seconds.

- Working memory:

Its role is to manipulate what is contained in short-term memory (working memory is considered to be part of short-term memory). In other words, it carries out the processing of information.

- Long-term memory:

Long-term memory allows us to remember our distant recollections. Long-term memory information is not accessible at any time. For example, we do not remember what we experienced at any time during our last vacation. On the other hand, with some effort, we can reconsolidate our recollections. The greater the number of contextual recall clues we have, the easier recall will be. We could compare this type of memory to a database in which we can access a former memory baggage by different means.

- Explicit memory:

Recollections that are part of explicit memory can be recalled consciously.

- Semantic memory:

It contains information about our general knowledge. For example, it allows us to remember the name of an object or the meaning of a specific term (key phrases for example).

- Episodic memory:

It contains information about events we have experienced. Unlike semantic memory, episodic memory is associated with a specific time and space (e.g., we do not know where and when we learned what "computer" means, but we know where and when we danced our first slow dance).

As regards episodic memory and semantic memory:

Memory appears as a great serial in which episodes merge to build up generic abstractions in the form of the words, faces, and places that are familiar to us.

Episodic memory is more fragile than semantic memory, so we often forget the name of a person or what happened when we went to such a place. Episodic memory can be retrospective or prospective:

• Retrospective memory:

It is simply a memory of past events.

• Prospective memory:

It allows us to remember future actions to be done. For example, remember that Wednesday at 8 a.m. we have an appointment at the dentist. This type of memory is associated with the frontal cortex since it involves planning actions. It is therefore very close to attention and requires a lot of mental resources.

• Implicit memory:

Implicit memory refers to things that one learns "unconsciously." For example, procedural memory is a system that allows us to learn new motor skills. Most people know how to ride a bike, but we do not need to consciously control our muscles in a certain way to succeed in riding a bike.

Many memories are "prodigious"; they are professional memories, e.g., chemists, historians, and literary people.

• Memory of smells:

Since the story of Proust's "madeleine," many people have thought that smells strongly recall memories (Lieury 2013, p.132).

4.4 How Does Memory Work?

Memory works mainly through three processes: encoding, storage or consolidation, and recovery.

First of all, encoding is the process that allows the brain to record information and thus forms so-called memory traces. Encoding is greatly influenced by attention and motivation. For example, if you are in the park absorbed in reading your new novel and one of your friends is walking past you, you most likely will not notice him/her. It is then impossible for you to create the memory of having seen your friend since you were too concentrated on the plot of your novel. On the other hand, if you were just taking a short break from reading and your friend passed by, you will probably notice him/her. That is when the encoding of this event became possible.

Once the piece of information is recorded in our brain, this memory trace has to become durable. Storage (or consolidation) transforms the memory trace so that it can be stored in memory. Consolidation, which is in fact a set of strategies for maintaining information on the long term, makes it possible to keep a strong mnemic record. For example, to remember having come across your friend in the park, your brain will have to make links with pre-existing information such as your recollections of your friend, of the park, etc. These recollections are valuable sources of information to properly integrate the event into your memory. In short, storage creates links between the various pieces of information we have in memory. This allows us to recall them more easily afterward. It is also important to remember that consolidation is a process that occurs in an unconscious and automatic way but can be promoted (by good study methods for example).

Subsequently, once the recollection has been properly recorded and remembered, one must of course be able to go and retrieve it. To access an event stored in long-term memory, the human brain uses a process called recovery. Recovery reactivates recollections. To do this, it is possible to use mental cues (e.g., think of the last time we saw our friend) or external clues, such as location, smell, etc. (e.g., going to the park can remind me that I came across a friend of mine) to help us remember what happened.

Among these three processes, encoding and recovery are easily affected by factors that can affect the memory of our recollections. As already mentioned, encoding is greatly influenced by attention, motivation, and also by the strategies our brain uses to store a recollection in memory. The same factors play a role during recovery. Consolidation is the only process uninfluenced by external factors, since it occurs more automatically. Since encoding and retrieval are the processes most affected by external factors, they are also the ones that become less effective with age (in the inner capsule, what is normal aging, mild cognitive impairment? Alzheimer's disease?). On the other hand, it should not be forgotten that, regardless of age, it is always possible to learn new strategies to properly encode or retrieve information. Here are some that may be interesting:

If you have a list of words to learn, you can create a simple sentence containing the words you need to memorize. To help you remember things to do or a list of words, you can also use the mental imagery strategy. For example, if you have to go and get a chocolate cake at the bakery and clean your fridge, you can create a mental image that includes both things to do (e.g., cake in the fridge). If you have trouble remembering a person's name, you can associate their name with an attribute or trait that strikes you in that person (e.g., this person has beautiful large green eyes).

It is important to note that the strategy of repeating many times in our head the words we need to remember is not the most effective strategy because this way of encoding information is very sensitive to interference.

4.5   Where is Memory Located?

For information, the brain contains about 86 billion neurons, including 16 in the cortex and 4 in the visual system. It is not possible to specify the exact location of memory in the brain, but we know that certain brain areas are involved in certain

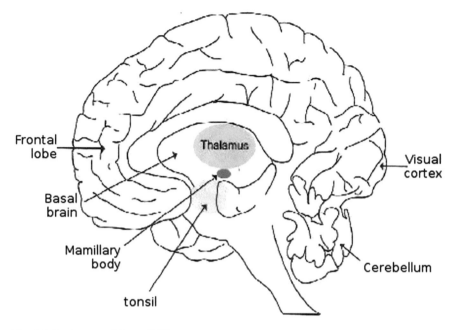

Fig. 4.1   Brain zones (Lieury 1992)

processes. Memory is said to be a multimodal function, which means that it uses several modalities and therefore several brain regions.

Here are the structures most involved in the operating of memory (Fig. 4.1):

- The front lobe:

It controls encoding and recovery. Encoding cannot be done without us (with the help of the frontal lobe) focusing our attention on the thing to be memorized. It is the frontal lobe that rakes our memory for recollections, thanks to multiple strategies. Finally, it allows us to make judgments and inhibit the recollections that are not useful at a precise moment (it decreases interference). The frontal lobe is the lobe that controls the entire brain.

- The temporal lobe:

It is used to store information in long-term memory. Unlike the frontal lobe that acts as a search engine, the temporal lobe is comparable to a database. These two brain regions are therefore indispensable and complementary to reactivate any recollection contained in our long-term memory.

- The hippocampus:

It acts as a memory recorder.

It consolidates memories. It transforms mnemic traces into short-term memory and into recollections in long-term memory. It is this structure that brings together the different parts of an event to form a complete recollection. The hippocampus plays a vital role in creating new recollections or learned items, and it consolidates information to store it in the cortex.

- The cerebellum:

It is the seat of automatisms. It may well also be the seat of procedural memory.

- The amygdala:

It plays a determining role in the consolidation of emotional recollections. This structure associates each recollection with the appropriate emotion, so that the recollection may be easier to recall later.

The American neurobiologist Joseph Ledoux (1994) showed that in the brain, near the hippocampus which is the recorder of our memory, another structure is attached: the amygdala. The amygdala is the emotional brain of recollections.

4.6   The Neurons of Memory

In his very recent article (October 2017), Alcino J. Silva analyzed how one recollection followed from another. He supported the following theory, " Observing recollections forming in the brain and even seeing two recollections being related is now possible thanks to new types of microscopes."

Our memories depend on our ability to remember the details of the world: a child's face, a duck, a lake...

The field of neuroscience is currently tackling a fundamental question: how does our brain connect our recollections with one another across time and space? So far, most of research about memory has focused on how we acquire, store, revive, and alter our individual recollections. But most recollections are not unique, isolated entities. We know that the recalling of one recollection revives another.

To study the way recollections are associated, the discovery of "neural allocation" was a determining step forward (Silva 2017). The authors found that the brain uses specific rules to assign fragments of new information to different groups of neurons in brain regions involved in building up memory.

In fact, around 1998, Michael Davis's team manipulated a gene named CREB in rats in order to amplify their emotional memory, e.g., the association of a beep and an electric shock.

In the past, Alcino J. Silva's laboratory (nowadays at the University of California in Los Angeles (UCLA)) had discovered, in collaboration with other researchers, that the CREB gene was necessary for long-term memory to form. It fulfills this function by encoding a protein that regulates the expression of other genes necessary for memory. During a learning process, certain synapses (the cellular structures that neurons use to communicate with one another) are formed or reinforced to facilitate exchanges among neurons. The CREB protein acts as a molecular architect in this process. Without it, most of the things we live would be forgotten. Michael Davis's team succeeded in improving memory by increasing CREB production in only a small subset of the amygdala neurons, a brain region essential for the memory of emotions.

They focused on the functions of the CREB protein in two key brain regions for memory: the abovementioned amygdala, but also the hippocampus, which is crucial in the memory of loci.

Sheena Josselyn used a virus to introduce additional copies of the CREB gene into mouse amygdala neurons. She showed that these neurons were almost four times more likely to store scary recollections than neighboring cells. Evidence has been accumulated that emotional recollections are not attributed to random neurons in the amygdala. The cells that upload recollections are the ones that produce the largest amounts of CREB protein.

The same team took advantage of a revolutionary technique called optogenetics, which uses blue light stimuli to activate or inactivate neurons. They showed that it was possible to artificially cause mice to recall a recollection of fear by using blue light to activate the amygdala neurons richer in CREB.

The ability of CREB to designate which neurons are part of a recollection—what they called "neural allocation"—led them to the hypothesis that this process could be the key to our ability to bind recollections to one another.

Denise Cai (2016) proposed a clever experiment: Justin Shobe and she placed the same mice in two different cages on the same day, five hours apart, in the hope that recollections of the two cages would be connected to one another. Later, they gave them a slight electric shock while they were in the second cage. As expected, when the mice were once again placed in the cage where they had received the electric shock, they remained motionless, probably because they remembered having received such a shock in this place. Remaining motionless is a natural reaction in frightened mice, because most predators detect them less easily that way.

Mark Schnitzer (Stanford University) described a tiny microscope that his lab had invented, able to visualize the activity of neurons in the brains of free-moving mice. This microscope, which weighed 2–3 g, was worn by the animal like a hat.

In older mice, recollections were less related because the neural networks that stored them overlapped less. But this link could be restored.

• Alcino J. Silva concluded with these words:

"Perhaps understanding these phenomena will one day help us develop treatments for memory disorders seen in diseases such as depression, age-related cognitive decline, schizophrenia or bipolar disorder ? One thing is certain: research on the formation of recollections is now less constrained by the range of available techniques than by the limits of our imagination."

These very recent research works published by American people make me feel as if I discovered science fiction.

4.7   More Neurons for a Tidier Memory

For centuries, it was believed that the adult brain could not make new neurons.

After all, as the brain stores information in specific networks of nerve connections, one might think that the random introduction of inexperienced cells into these networks would paralyze our ability to encode and correctly retrieve information, and thus spread confusion among our recollections.

By carefully examining brains from rodents, monkeys, and even adult humans, researchers showed that new neurons continue to appear throughout our lives in two brain regions: one is involved in smell, and the other, the hippocampus, is involved in learning, memory, and emotions.

Since then, neuroscientists have questioned the exact function of these neoneurons. And while their role in the olfactory system is still somewhat obscure, the hippocampus has begun to reveal its secrets.

The works of Kheirbeck and Hen's team (Kheirbeck and Hen 2017) and others suggest that newly formed neurons help record recollections in a way that individualizes them and prevents them from being superimposed. This discovery could lead to new approaches to treat anxiety disorders, including post-traumatic stress disorders, because people with these disorders have difficulty in describing the difference between true fear situations and danger-free situations.

Proust's madeleine

Basically, memory consists of roughly equal numbers of reminders of recollections and recordings. Most often, it is the first process whereby a precise and detailed recollection emerges at a glance, a smell or a taste that causes astonishment. The smell of a madeleine dipped in a cup of tea immediately brought back the narrator of Marcel Proust's novel "In Search of Lost Time" to the Sunday morning of his childhood:

"And as soon as I recognized the taste of the madeleine piece soaked in the lime tea that my aunt gave me ... the old gray house on the street, where her room was, came and applied itself to the small detached house like a theatrical scenery, overlooking the garden [...]; the whole of Combray and its surroundings, all of which took shape and solidity, came, town and gardens, out of my cup of tea."

The ability to remember a full, complete recollection from a perceptive clue (called pattern completion) is one of the most important functions of the hippocampus. However, for a memory to come back, it has to be properly deposited. Another fundamental task of the hippocampus is the ability to distinguish two memories with similar sensory inputs ("pattern separation"). Thanks to this ability, which appears to be related to the production of new neurons, we remember (in most cases) the place where we parked our car in the morning, as opposed to the one where we left it the day or the week before.

To test the ability of mice to distinguish between close recollections, they placed them in a very similar but different cage. For example, if the cage where they received the shock was square with silver-colored walls, blue lighting, and a clear smell of aniseed, the other cage had the same shape and color, but a smell of banana or lemon. First the animals were frightened. However, if no shock ensued, they quickly learned to distinguish between the two situations, stopping in the cage where they received the shock and showing no fear in the other.

The researchers' reasoning (Kheirbeck and Hen 2017) went as follows: if production of new neurons is essential to tell recollections apart, suppressing neurogenesis in the dentate gyrus of an animal should make it more difficult for it to distinguish between the two situations. And that was what they observed. The animals deprived of new neurons remained excessively nervous, reacting with concern in both environments, even after repeated passages in harmless cages. Unable to dissociate the two recollections, the animals generalized their fear of the original cage, extending their anxiety to any place that resembled it.

Please note that the dentate gyrus (gyrus dentatus) is a gyrus of the limbic lobe of the cerebral cortex. It is located above the hippocampal groove, along the hippocampus.

Conversely, researchers experimentally amplified the number of neo-neurons in the mouse dentate gyrus by removing a gene that usually promotes the death of any new nonessential cell. The mice whose dentate gyrus contained a greater density of new neurons distinguished the electric shock cage from the one that resembled it more easily, and they felt safe in the danger-free cage more quickly. These observations confirmed that neo-neurons play a role in encoding and in the ability to distinguish between similar but different recollections.

Neo-neurons to slow down brain activity

These studies were not conducted on humans. But if neurogenesis is important for us to distinguish between recollections, we may expect an interruption of this process to be associated with some kind of disruption of the activity of the dentate gyrus, where neo-neurons are formed and remain. In fact, such a link has been observed in human subjects. Using functional magnetic resonance imaging to detect neuronal activity, Michael Yassa, then at Johns Hopkins University, Baltimore (United States), and Craig Stark, of the University of California, Irvine, showed that people having difficulty in distinguishing different similar objects exhibited a strong activity in the dentate gyrus.

A method designed to increase the production of new neurons could act more specifically. A recent search for chemical compounds stimulating neurogenesis in the mouse dentate gyrus revealed a promising candidate, named P7C3, which promotes the survival of neo-neurons.

This research is quite recent (the publications were issued in 2017). As of now, it will be necessary to reinforce the results specifically obtained in men and women.

References

D.J. Cai, A shared neural ensemble links distinct contextual memories encoded close in time. Nature **534**, 115–1118 (2016)

B. Chamak, Dynamique d'un mouvement scientifique et intellectuel aux contours flous : les sciences cognitives:(etats-unis, France). Revue d'histoire des sciences humaines **25**(2), 13–33 (2011)

F. Dworczak, *Neurosciences de l'éducation: cerveau et apprentissage* (L'Harmattan, 2004)

J. Ecalle, A. Magnan, L'apport des sciences cognitives aux théories du développement cognitif: quel impact pour l'étude des apprentissages et de leurs troubles? Revue française de pédagogie **152**, 5–9 (2005)

M. Gaussel, C. Reverdy, Neurosciences et éducation: la bataille des cerveaux. Dossier d'actualité, veille et analyse **86**, 1–40 (2013)

N. Herzberg, Le prix nobel, science inexacte, in *Le Monde*, 3 Octobre 15 (2017)

M.A. Kheirbeck, R. Hen, Des neurones en plus pour une mémoire bien rangée. Pour La Sci **480**, 38–44 (2017)

A. Lieury, *La mémoire* (Mardaga, 1992)

A. Lieury, *Le Livre de la Mémoire* (Dunod, 2013)

J. Silva, Alcino. Comment un souvenir en appelle un autre. Pour La Sci **480**, 28–38 (2017)

G. Tiberghiem, *Dictionnaire des sciences cognitives* (Armand Colin, 2002)

J. Valera, Francisco, in *Invitation aux sciences cognitives* (Editions du Seuil, Paris, 1997)

Chapter 5
A Philosophical Approach to Memory

Abstract At first, there is memory of thought, a construction of the past and also models for thought and storage. Memory works in networks. Memorization results from a modification of connections among neurons. One speaks of "synaptic plasticity."

The history of philosophy teaches us that there are epistemological problems concerning memory (Bollack 2004). Where is the memory-image when we no longer think about it? How do I find an image from the past? How can it be in the past and reappear in the present? As for forgetfulness, why do some images or recollections fade away while others remain? Where do they go? Where do these recollections spring from when they are released by psychoanalysis, for example? As regards history, what are the differences between history and fiction, a history book and a novel? How do historians establish the truth of their claims? Does history have an "objective" truth, or are there many points of view about any event? (Reagan 2008).

5.1 Memory of Thought

Greek and Latin roots: memory and "mental"

Memory may not exist per se, as a distinct intellectual faculty.

Thinking is connected with what it knows, which we must not lose sight of, or which we might not think about; we might "forget."

Memory has attributed itself a proper ground in speech, which it is difficult not to relate to the cultural importance of past recollections.

Strength and delirium:

Homer showed that memory in action was at the service of social order.

The rule is reminded to avoid forgetfulness from within and escape caused by excess from without.

Memory is linked to the exhortation and upgrading of social values, forgetfulness to their nonrecognition.

© Springer Nature Switzerland AG 2019

C. Lexcellent, *Human Memory and Material Memory*,

https://doi.org/10.1007/978-3-319-99543-4_5

The two kinds of forgetfulness: too much or not enough intensity

1. A boundless thought
   Memory, a creator of values, is implicitly defined through contradiction.
2. The perils of forgetfulness
   But forgetfulness can occur, not from excess intensity, but from a defect in the maintenance of thought.
   The psychological and gnoseological question is to know how memory works. Plato chose the writing model, and later Freud did so too.

5.2 A Construction of the Past

The constructions of history and the war of memories

The past is both elusive and available, and hence the war of memories.

Nietzsche: an ontologization of fate

Nietzsche circumscribed an aporia: man is condemned to flee the oblivion of childhood; he is bound to the knowledge of his past although it crushes him.

Only the high concentrations of life are important. He speaks of "the creative force of an individual, a people, a civilization" to embrace a globality. Science is nothing in itself, it is intended for power, past reigns serve the reigns to come. Memory is rehabilitated as a place of reincarnation or resurrection. Memory is saved when brightness is found in this other messianism made of the cyclical culminations of life. The remembered life of history is reaffirmed in the blossoming of life.

Memory relates to what is imposed: nothing inferior ever and no projection into the future.

Poetic memory: Mnemosyne and Lesmosyne

Memory is the mother of the Muses. Immortalization comes second. The construction of the past, extended to the gathering of all human knowledge, was transformed in ancient times when it was entrusted to specialists of actualization through words and music. Memory thus became the business of a profession. The function (memory?) was related to festivals and celebrations, and a new science opened up onto inventions and reinventions of multiplied horizons.

Forgetfulness is not the absence of memory or its erasure, but more positively, a wrenching away from the avatars of an ordinary, alienating existence.

The fiction of an entirety of knowledge

1. Homer's Muses
   The Muses knew everything, they represented the abstraction of an all-powerful art. Entirety in space or time, in the world or in history, was part of an unlimited superhuman memory.

2. Beaudelaire's Andromache: pain as a muse

"The immense majesty" of the widow's tears is echoed by the fertility of an already fertile "memory," fertile like the earth, containing all that could ever be said and written later.

3. Mallarmé's cutoff: freedom through forgetfulness

Forgetfulness is the first condition of poetic creation.

5.3 Models for Thought

Writing

1. Plato: memory and knowledge

Men are variously marked with a seal.

"Ideas" or impressions that form admittedly leave an imprint inside us, on something believed to be a block of wax that we carry within us.

Forgetfulness occurs when the inscriptions are erased and lost; inequality among men is explained by the volume and quality of the impregnable mass that each of us possesses (it is the same for certain metals!).

A piece of knowledge disappears if one considers that there is a false opinion and "that one is able to not know what one knows."

The wax memory is the "core" of the soul.

The magic of innumerable fixations is indeed a gift of memory to the human soul.

2. Freud: writing in the unconscious

The object of the reconstruction of memory by psychoanalytic therapy exists in a written form. In Freud's eyes, a language was originally engraved in the body, at an age when the infant was driven by fits of its impulses. It bears their marks. The deviations and accidents of this natural history turn the body into a text that can or cannot be read.

The basic principle is to stringently separate consciousness from this buried memory made of successive inscriptions.

A partially intelligible memory opens access to the records or archives of the unconscious, and it allows for a gradual interpretation of amnesia through recollections.

3. Bergson: traces of our life experiences in our involuntary memory

In his chapter on "The two forms of memory" ("Matter and Memory", p. 80 Bergson 1896), Bergson highlighted a memory that escapes representations, according to the model of a learned recollection. It reveals itself as a rationalistic prejudice, an ideology hindering the perception of this other, "spontaneous" and immediately perfect recollection. What counts is "the recollection that must not become alien to our past life," be alienated by all the rest, by external influences.

Bergson singled out spontaneous memory, distinct from remembering, devoted to the learning of the mechanisms essential to our role as social actors.

4. Benjamin: memory of strata

Walter Benjamin valued Proust's search by adding a transcendental aspect into it. Through recollections, our self discovers its own existence by recognizing the stages of awareness of what it does. The search for our ego teaches us that what we remember is the image of the act that marks it out, and that this image is the result of the "thought" that engages in it.

5.3.1   Storage

1. The German language, between updating the past and hoarding

The faculty to remember appears to be multiplied in this place of recovery where all that has been thought is gathered.

2. Hegel: the concept and the penetration of its associative potentialities

Internalization encompasses traditional self-examination and the whole dimension of self-reflection.

3. Poetic memory integrated into the philosophical system

The function of memory leads to the ripping of poetry.

4. Heidegger's split internalization

What one "remembers" is not called upon by consciousness but suddenly emerges and imposes itself. From a more theological point of view, the reception of fundamental truth defines the authentic existence of the self; memory will therefore consist in "letting fundamental truth come to meet us more and more." Memory is set within the framework of contemplation.

5. Heidegger's Holderdin: truth concentrated in the words of memory

For Paul Celan, the word "memory" is alive: it expresses itself in its own condensation, and it speaks to itself.

5.4   Memory According to the INSERM

Memory makes it possible to record information coming from experiences and various events, preserve them, and restore them. Different neural networks are involved in different types of storage. Better knowledge of these processes improves the understanding of certain memory disorders and opens onto possible future interventions.

Visual-spatial memory (and also pictorial memory)

It goes back to the legendary discovery of the method of loci (see Simonides of Ceos). Special areas of the brain are involved in spatial memory, including the parietal cortex and the right hippocampus.

Memory relies on five memory systems

This representation of neuroimaging is an example of the technique known as interindividual synchronization guided by grooves (Diffeomorphic Sulcal-based COrtical or DISCO) © Inserm, G. Auzias/S. Baillet/O. Colliot.

Working memory

Working memory (or short-term memory) is actually the memory of present time. It allows us to retain information for a few seconds, even a few tens of seconds. We constantly request it at every moment, for example, to remember a phone number long enough to note it. In most cases, the neurobiological mechanisms associated with working memory do not allow for long-term storage of this type of information: their recollection is quickly forgotten. Nevertheless, there exist interactions between the working memory system and long-term memory systems. They allow us to memorize certain events and thus call on old recollections in the face of present situations in order to better adapt ourselves.

7, the magical number

The number 7 is believed to be the "magical number" of working memory. It is the number of elements that can be memorized simultaneously in the short-term, plus or minus two events. We are all able to remember between 5 and 9 items for 5 s on average. For example, the sequence [7, 9, 6, 4, 0, 9, 2] represents 7 digits. It can also be read [796, 409, 2] which represents only three (and leaves the possibility to remember four other items). Similarly, a series of long and complicated words like [parrot, hummingbird, spider, Diplodocus, chimpanzee, kangaroo, platypus] represents 7 words that can be remembered, although it is composed of a much larger number of letters. Various mnemonic processes use this property of our brain to expand the capabilities of working memory.

Memory consists of five memory systems involving distinct but interconnected neural networks:

Working (short-term) memory is at the core of the network.

Semantic memory and episodic memory are two systems allowing for long-term conscious representations.

Procedural memory allows for unconscious automatisms.

Perceptive memory is related to our senses.

This complex set is indispensable to identity, expression, knowledge, awareness, thinking, and even to our projection in the future.

Semantic memory

Semantic memory allows us to acquire general knowledge about ourselves (our history, our personality) and the world (geography, politics, news, nature, social relationships, or professional experience). It is the memory of awareness and knowledge. It concerns personal data that is accessible to our conscience and that we can express.

Episodic memory

Episodic memory is a form of explicit memory. It allows us to remember past moments (autobiographical events) and to predict the next day. When someone is asked to evoke the recollection of an event that took place in the last few months or to think about the next vacation in order to imagine what will happen, the same brain circuits are activated. The details of episodic recollections are lost over time (where, when, and how did the event happen?). The traits common to the different lived events mingle with one another to become knowledge that is no longer related to a particular event. In the end, most episodic recollections are transformed into general knowledge.

Procedural memory

Procedural memory is the memory of automatisms. It allows us to drive, walk, bike, or ski without re-learning every time. This type of memory is particularly called on by artists or athletes to acquire perfect procedures and reach excellence. These processes are performed implicitly, i.e., unconsciously. People cannot really explain how they do it, why they keep their balance on skis or go down without falling. Movements are done without any conscious control and the corresponding neural circuits are automated.

Perceptive memory

Perceptive memory depends on sensory modalities, including sight for the human species. This memory works a lot without individuals knowing. It allows us to retain images or noises without realizing it. It allows people to return home out of sheer habit, thanks to visual cues. This memory allows us to remember faces, voices, and places.

5.5   Memory Works in Networks

From a neurological point of view, there is not "one" memory center in the brain. The different memory systems involve distinct neural networks, observable by medical imaging during memory tasks or recovery of various pieces of information. These networks are nevertheless interconnected and work in close collaboration: a same event can have semantic and episodic contents and the same information can be represented in explicit and implicit forms.

Thanks to the two writing systems that the Japanese possess, their neurologists discovered that the ideographic language is stored in the right brain, while phonetic language is generally privileged by the left brain (Lieury 2013, p. 192).

Lateral (left) and internal (right) sides of the right cerebral hemisphere (see Fig. 4.1).

Procedural memory recruits neural networks in the subcortical area and in the cerebellum.

Semantic memory involves neural networks disseminated in very large regions as well as in the temporal lobes, especially in their most anterior parts.

Episodic memory uses neural networks in the hippocampus and more broadly in the internal face of the temporal lobes.

Finally, perceptive memory recruits neural networks in different cortical regions, near sensory areas.

Reasoning arises from multiple recollections

Memories rely on one another! If you know that a 4WD is a car, you can say that a 4WD has brakes, even if no one has told you and you have never seen one. You deduce this from the fact that all cars have brakes. This type of reasoning is useful in everyday life and is essentially based on the items of knowledge stored in our memory. Thus, the more numerous the items of stored knowledge, the easier it is to make analogies.

5.6 Encoding and Storage of Information, a Case of Synaptic Plasticity

The activation of the hippocampus is maintained for episodic recollections, but declines when recollections become semantic (© Inserm, C. Harand).

Storage processes are difficult to observe by brain imaging because they are part of long-term consolidation mechanisms. Nevertheless, the hippocampus seems to play a central role in the temporary and more durable storage of explicit information, in connection with different cortical structures.

Memorization results from a modification of connections among neurons of a given memory system: one speaks of "synaptic plasticity" or perhaps of "synaptic viscoplasticity"! (synapses are the points of contact between neurons). When information reaches a neuron, proteins are produced and routed to synapses to reinforce them or create new ones. This produces a specific network of neurons associated with the recollection that is engraved in the cortex. Each recollection therefore corresponds to a unique configuration of spatiotemporal activity of interconnected neurons. Representations end up being distributed across vast, extremely complex networks of neurons.

The regular and repeated activation of these networks is believed to strengthen or reduce these connections in a next stage, and to consequently consolidate the recollection or on the contrary forget it.

Sleep consolidates memory

A lesson is best learned in the evening before sleeping, it is a fact! Experiments on information recall show that sleeping improves memory, all the more so as sleep lasts long. On the other hand, sleep deprivation (less than four or five hours a night) is associated with memory problems and learning difficulties. Besides, 0.75 Hz electrical stimulations of the brain during the slow sleep phase (characterized by the recording of slow cortical waves on the encephalogram) improve the capacity to memorize a list of words. Several hypotheses could explain this phenomenon: during sleep, the hippocampus is at rest, and this is thought to avoid interference with other information while the recollection is being encoded. It could also be that sleep sorts out events, removing the emotional component from recollections to retain only the informational one to facilitate encoding.

Memory

File produced by Prof. Francis Eustache, Director of the Inserm-EPHE-UCBN Unit U1077 "Neuro-psychology and Functional Neuro-anatomy of Human Memory"— October 2014 (Eustache et al. 2014).

Memory makes it possible to record information coming from experiences and various events, to preserve them and to restore them. Different neural networks are involved in different types of storage. Better knowledge of these processes improves the understanding of certain memory disorders and opens onto possible future interventions. Memory is based on five systems of repeated memory of these networks and is believed to strengthen or reduce these connections in a second step. As a consequence, the recollection is consolidated or on the contrary forgotten. It is important to specify that forgetfulness is associated with the good functioning of memory, pathological cases excluded. Some works suggest the role of a molecule called PKM zeta in maintaining long-term memory. In animals, it helps maintain modified molecules during encoding and prevents their degradation over time, consolidating networks associated with recollections.

Long-term potentiation is probably the starting mechanism of the learning machinery at the level of synapses, the switches between neurons. The activation of a neuron results in the release of a neurotransmitter such as glutamate. Glutamate serves as a kind of key to open the "lock" of the dendrite (an extension of the neuron input) of the neighboring neuron. A receptor named NMDA is thought to play the role of "memory" at the level of the neuron. When the glutamate key unlocks the NMDA lock, the receiver valve opens and lets magnesium ions out, while calcium ions rush into the neuron. The calcium ion is a messenger that activates a cascade of enzymes and thus allows for receptors to remain open, and in turn for synaptic activity (Lieury 2013, p. 198).

The release of neurotransmitters, including glutamate and NMDA, as well as the expression of a protein that increases glutamate release, syntaxin, is associated with synaptic plasticity. Morphologically speaking, this plasticity is associated with changes in neural networks: changes in the shape and size of synapses, transformation of silent synapses into active synapses, and growth of new synapses.

Brain plasticity is significantly better before the age of 10, at least for motor areas.

During aging, the plasticity of synapses decreases, and changes in connections are more ephemeral, which may explain increasing difficulties in remembering information. Besides, in the rare forms of Alzheimer's disease that hit whole families, mutations are associated with plasticity defects of synapses that could explain major memory disorders.

Researchers are gradually discovering factors that increase memory capacities and seem to stabilize recollections over time. This is the case of the cognitive reserve, a phenomenon associated with extremely numerous functional connections between neurons resulting from learning, from lifelong intellectual stimulation, or even from flourishing social relationships.

Researchers do not presently know precisely which educational and social ingredients are involved in the formation of this cognitive reserve. Studies in rodents, however, showed that when animals stay in complex (so-called "enriched") environments, their ability to learn and memorize improves. Other studies conducted on humans indicated that people with a high degree of education develop the symptoms of Alzheimer's disease later than those who have not been to college or the like. These results, stemming from epidemiological research involving large numbers of subjects, could be explained by the brain's ability to compensate for neuronal degeneration linked to the disease through the mobilization of alternative circuits, thanks to a better network of connections among neurons in people who have a high level of education.

Other factors contribute to the consolidation of memory, but their mechanisms are not perfectly known yet: sleep (see above), physical activity, or good cardiovascular health. In general, a healthy lifestyle (sleep, a balanced diet, and physical activity) contributes to good memorization capacities.

In the collective book by Eustache et al. (2017), Robert Jaffard discusses animal memory from the point of view of social learning, which is widespread among different animal species that also transmit traditions. In this very interesting chapter, we discover that chimpanzees always use their wand to fetch termites and ants in tree trunks, or that adult meerkats (small African carnivores) gradually teach their young how to eat scorpions. The author mentions "left a new hunting technique in humpback whales, which consists in giving great caudal fin blows to knock out the fish."

In the same book, Catherine Thomas-Antérion describes the neural networks involved in face recognition. Her description of the parts of the brain involved in this recognition is very technical. She emphasizes that the loss of the identification of the other person's face (prosopagnosia) or the illusion of the look-alike is at the core of disorders that affect perception, memory, emotions, and control functions altogether. We can note that in Capgras delusion, a very close person (often the spouse) is taken for a look-alike or an impostor.

In Chap. 3, Hélène Amieva teaches us about society, the representations, and stereotypes; it conveys that can strongly influence the functioning of our memory with positive aspects (e.g., good social relationships) and negative ones (e.g., ageism).

For Jean-Gabriel Ganascia (Chap. 5), "left the understanding of the functioning of individual and collective memories has proved decisive for making systems evolve in depth, increasing their power and their adaptation to the surrounding world."

Finally, Bernard Stiegler (Chap. 6) wonders about the objects that build social memory (the Internet, computers, …).

References

H. Bergson, *Matière et Mémoire. Essai sur la relation du corps à l'esprit* (1896)

J. Bollack, *Vocabulaire Européen des Philosophies: MEMOIRE/oubli* (2004)

F. Eustache, H. Amieva, C. Thomas-Anterion, J.G. Ganascia, R. Jaffard, C. Peschanski, B. Stiagler, *Ma mémoire et les autres*. ESSAI LE POMMIER! (2017)

F. Eustache, J. Ganzscia, R.D. Jaffard, C. Peschanski, B. Stiegler, *Mémoire et oubli* (Le Pommier, 2014)

A. Lieury, *Le Livre de la Mémoire* (Dunod, 2013)

C. Reagan, Réflexions sur l'ouvrage de Paul Ricoeur: la mémoire, l'histoire, l'oubli. Transversalités **106**, 165–176 (2008)

Chapter 6
Materials Have Memory

Abstract Some materials are called "shape memory alloys." The physical key of "shape memory" consists in a phase transformation between a parent phase called austenite and a produced phase called martensite. If a specific thermomechanical treatment called "training" is applied, it can memorize two geometric shapes ("high" and "low" temperatures).

6.1   A Short Introductory Dialog

Stent catheter is shown in Fig. 6.1.

An introductory dialog by Professor Cogitus Martin:

- Some materials have memory !!!:

"Memory? They have recollections?"

"Yes! They are able to remember the shape they once had."

"So some materials are endowed with a kind of intelligence?"

"Somehow, yes ... What if I tell you that certain metals can have an intelligent behavior!"

"Impossible!!! ... I have a skull with a brain, the seat of my thought. Thanks to it, I can speak, I can think, I can decide, I can react, I can imagine, I can dream, I can remember, ... In short, I can say that I think. But metals cannot remember anything."

"Yes, they can: many metals can be deformed almost to perfection: they can be twisted or compressed or stretched without breaking, for example, and then they can return to their original shape without you making the slightest physical effort, as if this initial shape was memorized to be reproduced at the right time. Strange, isn't it?"

"What's the importance of these materials?"

"Multiple and essential since they are present everywhere around us, often without us knowing. They are found in medicine, dentistry, aeronautics, ... or more simply on the tip of your nose or your head ... Hey, glasses or hats that have memory? You could soon look at the world around you with different eyes..."

© Springer Nature Switzerland AG 2019

C. Lexcellent, *Human Memory and Material Memory*,

https://doi.org/10.1007/978-3-319-99543-4_6

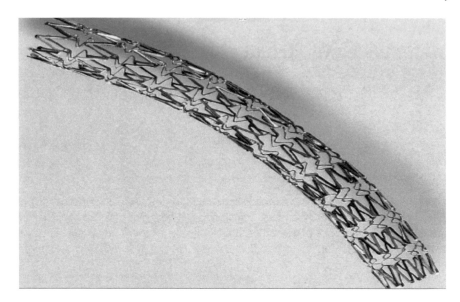

Fig. 6.1 A pseudoelastic stent (Lexcellent 2013)

A small physics experiment

You are at the bottom of a basin whose walls are densely covered with obstacles made of often inextricable coppice.

You want to walk up to one edge of the basin. The first time it is exhausting, you have to overcome all the obstacles that are in the way, i.e., tread over trees, bushes, brambles... You go down more easily. The second time you go up, it is already less painful; you start building a path. At the same time, it becomes almost easy, the passage is open. The higher you climb, the less it costs you in energy. We could say that you have educated nature or educated yourself!

If you come back 1 year later, the obstacles have built up again, you have to start it all over again. You have the impression that you have forgotten all the work done a year ago, in short to have become amnesic!

This little story can be applied to a cyclically solicited metal alloy.

6.2   Martensitic Transformation

What are these alloys known as shape memory alloys (SMAs)?

First, they are metal alloys with two, three, or even four components, and with very special compositions.

There are mainly two families:

- Cu–Al copper-based materials (Zn, Ni, Be, etc.);
- nickel-titanium-X (with X in low proportions) Ni–Ti (Fe, Cu, Co, etc.).

These materials are called "memory-endowed," that is, they have the property to "remember" the thermomechanical treatments they have been subjected to (in tension, torsion, flexion, etc.).

The geometrical shapes they had at low and high temperatures are two states whose memory they can keep. This memory was developed through a training treatment, that is to say often by the repetition of a same thermomechanical loading made of a same sequence of constraint (force) or imposed deformation (displacement) or (and) temperature.

The physical key of "shape memory" consists of a phase transformation between a parent phase called austenite A and a produced phase called martensite M. This phase transformation is termed thermoelastic for SMAs. It is a change in the crystal lattice between a "high temperature" phase A and a "low temperature" phase M, called "martensitic transformation." Austenite is transformed into variants of martensite. In the graphs below, in planar representation, i.e., in two dimensions, austenite is represented by a square whose side is a_0 and a variant of martensite by a rectangle whose sides are a and c and another variant whose sides are c and a (Fig. 6.2).

As shown by Chu and James's photographs of copper-based SMAs (Figs. 6.3 and 6.4) (Bhattacharya 2003), the microstructure can be very complex, making its analysis tricky.

As a reminder, a microstructure is what is generally observed under a light microscope, a nanostructure under an electron microscope, and finally a macrostructure is observed with the naked eye.

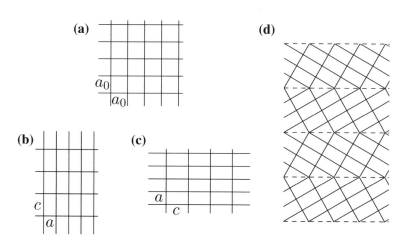

Fig. 6.2 A schematic illustration of martensitic transformation: **a** (a_0) austenite; **b** and **c** (a, c) martensite variants, **d** a coherent arrangement between martensite variants (Bhattacharya 2003)

Fig. 6.3 Optical micrograph
of the microstructure of a
complex-lattice Cu–Al–Ni
alloy, horizontal extent 0.75
mm. Courtesy of. Chu and
R.D. James (Bhattacharya
2003)

Fig. 6.4 Optical micrograph
of the microstructure of the
complex network of a
"corner-type microstructure"
Cu–Al–Ni alloy, horizontal
extent 0.75 mm. Courtesy
of. Chu and R.D. James
(Bhattacharya 2003)

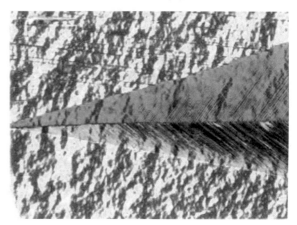

6.3   Two-Way Memory Effect: Training

If a specific thermomechanical treatment called "training treatment" is applied to an
SMA, it can memorize two geometric shapes, one in the austenitic phase, and the
other in the martensitic phase, under mere thermal loading. Similarly to educating,
one often repeats again and again! The treatments often consist of cyclic loadings:
for example, thermal loading. Two-way memory effect: training cycles under a constant force. The shape stored in the martensitic phase "results from the preferential
formation," in the absence of any applied macroscopic stress, of variants oriented
by the field of internal stresses generated in the material by the training process
(Patoor et al. 1987).

This can be schematized in Frémond's approach (Fremond 1998, 2002); in a
schematic model with two martensite variants, M_1 of volume fraction z_1 and M_2 of
volume fraction z_2, training boils down to "cutting a corner of the triangle"!

For virgin material, the martensite domain covers the entire triangle C_r (Fig. 6.5).

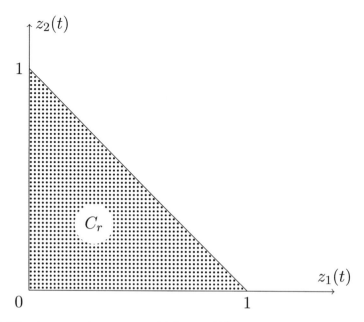

Fig. 6.5 Triangle representing a raw material (Fremond 2002)

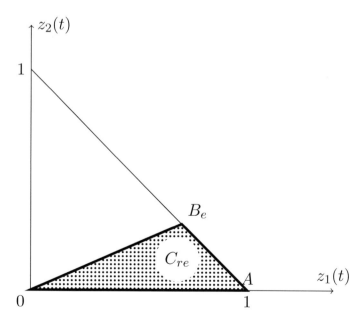

Fig. 6.6 Triangle representing a trained material (Fremond 2002)

For the trained material, the domain of existence of trained martensite lies in the triangle C_{re} (Fig. 6.6). This means that any point of the triangle C_r is a priori accessible first for virgin material. On the trained material, only the triangle C_{re} remains accessible.

6.4 Acquisition of Memory by SMA Training

A polycrystalline Cu–Zn–Al SMA manufactured by Tréfimétaux was used (Lexcellent et al. 2000).

First, a thermal treatment called "betatization" was applied for this ternary alloy to become a shape memory alloy.

Characteristic phase transformation temperatures (parent phase: austenite A; produced phase: martensite M) were measured in a state free from external mechanical stress, with sigma constraint = 0).

$$M_F^0 = 30\,°C, M_S^0 = 40\,°C, A_S^0 = 42\,°C, A_F^0 = 52\,C$$

$T\,(K) = 273 + T\,(°C)$, where $T\,(K)$ is the thermodynamical temperature.

The training process consists in applying a tensile constraint on the specimen in the austenitic state $T = A_F^0 + 40\,°C$ and the stress is kept constant for 10 cycles (Fig. 6.7).

Please note that 1 MPa $= 1 N/mm^2$. N represents the force expressed in Newton.

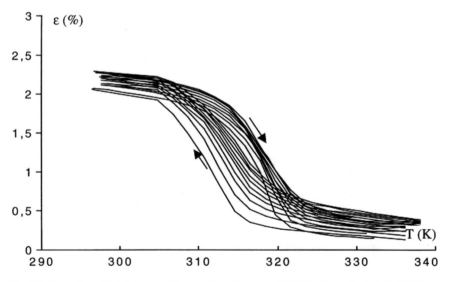

Fig. 6.7 Evolution of total strain ($\varepsilon\%$) over 10 cycles; $\sigma = 49\ MPa$ (Lexcellent et al. 2000)

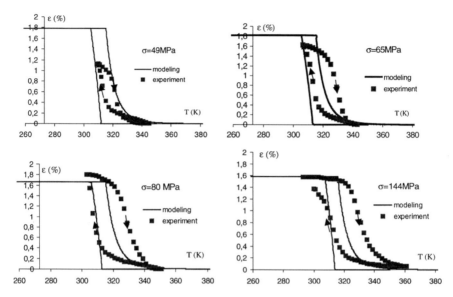

Fig. 6.8 Modeling of a two-way shape memory effect following the application of different stress levels (Lexcellent et al. 2000)

Then, the mechanical stress is removed. A purely thermal cycle is carried out, i.e., without any mechanical stress. Figure 6.8 shows the elongations of the specimen, obtained after cycling at different levels of stress ($\sigma = 49, 65, 80, 144$ MPa). A "training output" can be estimated as the ratio of the elongation value after training over the maximum elongation value obtained during training. We can note that this training yield saturates with the level of stress applied during thermomechanical loading!!! This means that educational success is not proportional to the load (effective, emotional !!) but many educators already know about that!

Perkins and Sponholz (1984) analyzed the mechanism of the two-way memory effect (or training). The obvious question was to find out if there were any physical characteristics (i.e., traces) in the microstructure associated with training. For example, the successive cycles might induce dislocation substructures (linear defects in the crystalline lattices) which act as nucleation points for the "preferred" martensite. In order to capture the microstructural changes that go along with training, a transmission electron microscopy observation program was started.

The microstructure of the virgin (i.e., not mechanically stressed) alloy is shown in Fig. 6.9. The austenitic parent phase contained no residual martensite or dislocation substructure.

All the microstructures were observed in the austenitic state, i.e., after removing mechanical loading (0, 3, and 15 cycles).

The substructure developed after a few loading cycles is illustrated in Fig. 6.10. Significant changes are clearly visible. An apparent increase in dislocation density is seen in the form of dark bands. When returning to the austenitic state, remnants of

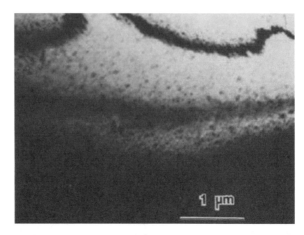

Fig. 6.9   Austenite microstructure in a stress-free state (Perkins and Sponholz 1984)

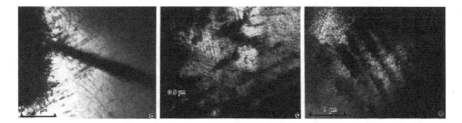

Fig. 6.10   Austenite microstructure after three cycles (Perkins and Sponholz 1984)

these bands in the form of dislocation entanglements and residual stress/constraint fields can still be seen.

After 15 cycles corresponding to training saturation, dislocation areas are clearly visible in the micrograph in Fig. 6.11. It is now evident that thermomechanical loading (the training process) has left traces (a memory imprint) in the material.

The memory of water

The memory of water is the name given, during a media controversy in 1988, to a hypothesis by the researcher and immunologist Jacques Benveniste, stating that when water has been in contact with certain substances, it keeps an imprint of certain properties of these substances even though they are statistically no longer there.

A series of experiments conducted to validate this hypothesis was then presented by advocates of homeopathy (who use very important dilutions of the active principles) as a scientific validation of that theory.

English researchers attempted to replicate the experiment, but they obtained negative results, and no satisfactory explanation was proposed, so chemists considered that the concept of the memory of water was only an experimental artifact. However, it is still studied by a few scientists, including Professor Luc Montagnier, laureate of

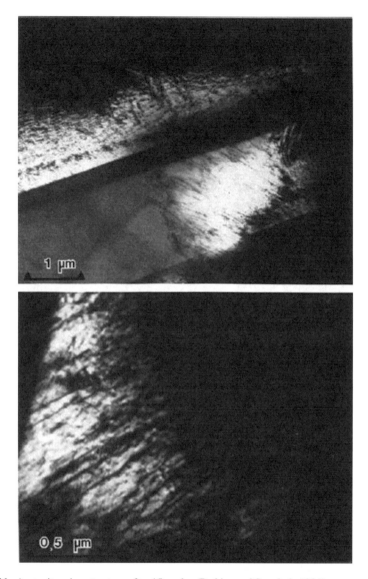

Fig. 6.11 Austenite microstructure after 15 cycles (Perkins and Sponholz 1984)

the Nobel Prize in Medicine in 2008, who believes that Benveniste was overall right, despite results that "were not 100% reproducible."

The scientific community admits as a principle that liquid water does not retain ordered networks of molecules for more than a small fraction of a nanosecond. In other words, no "memory traces" are created resembling dislocation structures in shape memory alloys.

A "magic" approach of shape memory alloys

When I started the thermomechanical study of shape memory alloys in 1988, during Pierre Vacher's thesis (he is now a professor at the IUT of Annecy), we established contacts with Gérard Guénin (then a professor at the INSA of Lyons) and also Alain Dubertret (a CNRS researcher).

When Alain Dubertret offered the president of Souriau company (a leader in connectivity appliances) to give him the secrets of shape memory metals, he subjected his collaboration to an unexpected condition, i.e., to let him work as he wished with the artistic world. He did not invent shape memory metals. Nobody knows exactly to whom we owe this miraculous finding, thanks to which a metal worked on for a long time can keep the memory of a shape and find it again when exposed to a certain temperature. The Americans have been using shape memory for several years, but others used it before them. "Hairstylists, when they sculpt the hair during a perm, do exactly what we do with metals to train them," he explained. What Alain Dubertret invented was the memorization of two shapes, for which he holds a patent with two other researchers and the Souriau company. On either side of a transition temperature, the object positions itself in two very specific shapes. Immediate and innumerable applications are possible for artists: sculptures could spring under the mere effect of a jet of cold water or change appearances between day and night, lampposts could emerge when temperatures go down, etc. From a less bucolic point of view, industrialists have other ideas: connectors can be plugged with zero effort at the temperature of liquid nitrogen to tighten themselves once heated. The military appears to be crazy about this device for their missiles. Besides, Alain Dubertret has other projects. First, he would like to find out, if possible, how to memorize several shapes in a same object. Then, he would like to act on the alloy through other phenomena than temperature changes (electromagnetic fields, for example). "Thus I would have several words available to talk to them...".

In the wake, Alain Dubertret and Jean Benveniste met to confront SMA memory and memory of water: Their meeting was reported in the magazine Explora.

It should be noted that Ni–Mn–Ga shape memory alloys, also called magnetic SMAs, can be actuated by a magnetic field (Lexcellent 2013).

6.5   Other Memory Systems

IBM is currently developing memory-based electrical resistors called "memristors."

The idea of inventing a system capable of demonstrating intelligence by imitating, no more and no less, the functioning of the human brain could well move from science fiction to science itself! If this feat became reality, it would be, among others, thanks to the research conducted by a team from the University of Kiel (Germany) and a mysterious electronic component called a memristor.

In the human brain, neurons are interconnected through synapses, and the whole system forms an incredibly complex network. Throughout our lives, as we learn, synaptic connections created one day are reorganized to fit to new situations. Scientists have named this phenomenon "synaptic plasticity."

But what makes our learning and memorizing ability effective is the fact that our brain is able, in parallel, to synchronize brain waves, the same ones that transmit neuron-to-neuron information. It is these two phenomena, of paramount importance in the functioning and efficiency of our brain, that researchers succeeded in reproducing at the core of an electronic circuit.

To that end, they used an electronic component of a new kind, a memristor. The principle of the memristor—for memory resistor, understand "resistance to memory"—was born in the 1970s. But it was only in 2008 that it was physically implemented for the first time. What is it about? A component that acquires some resistance when crossed by a current and, more surprisingly, retains a certain memory of the flow of charges that it may have received in the past, even after being disconnected.

Thanks to memristors, engineers could develop a new type of memory for our computers. Current memories are unable to retain information once the computer is turned off. At restart, data recovery on the magnetic disk is more or less time- and energy-consuming. Memristors could help avoid this step to gain in efficiency.

Researchers at Kiel University, for their part, sought to use memristors to mimic the functioning of the human brain. They connected two oscillators with each other using memristors. Oscillators, a little like what neurons do, produce periodic electrical signals. Researchers were able to observe that these signals, though originally independent of each other, gradually became synchronized ... as can be observed in our brain!

MICROPROCESSORS. On April 3, 2017, a French team (CNRS, Thales, Universities of Bordeaux, Paris-Sud and Evry) published an article in Nature Communications (Boyn et al. 2017), describing an artificial synapse (an electronic element called a memristor) which could well revolutionize deep learning.

To better understand the scope of the discovery, "Sciences et Avenir" spoke with Vincent Garcia, a nanoscience researcher and co-author of the study.

"Science et Avenir", "Can you specify what a memristor (contraction of 'memory' and 'resistor') is?"

Vincent Garcia, "It is a variable analog resistor that keeps in memory the electrical tensions applied to it. Technically, a thin ferroelectric layer is interposed between two conductors, and behaves like an electric dipole. The different physical domains of this material keep a memory of the applied electric field, and in particular in the intermediate states between two electrical pulses, where certain zones are positively charged and others negatively. It is typically this dynamic behavior that we studied and modeled at the scale of a memristor equipped with a thin ferroelectric ferrite bismuth layer. We then simulated the unsupervised training of an artificial network of 45 memristors for the recognition of simple shapes."

"Why can this component revolutionize the state of the art of deep learning?"

"We actually reproduced a synaptic function of the biological brain called Spike-timing-dependent plasticity (STDP). This function allows our artificial synapses to deal with forms of unsupervised learning, without consuming a lot of energy since the component 'remembers' somehow the history of the electrical impulses to which it was previously submitted."

In short, we can say that an SMA can function as a simplified brain, and a network of memristors too!

References

K. Bhattacharya, *Microstructure of martensite* (Oxford materials, 2003)

S. Boyn, J. Grollier, G. Lecerf, B. Xu, N. Locatelli, S. Fusil, S. Girod, C. Carrétéro, K. Garcia, S. Xavier, J. Tomas, L. Bellaiche, M. Bibes, A. Barthélémy, S. Saighi, V. Garcia, Learning through ferroelectric domain dynamics in solid-state synapses. Nat. Commun. (2017)

M. Fremond, L'éducation des matériaux à mémoire de forme. Revue Européenne des Eléments Finis **7**(8), 35–46 (1998)

M. Fremond, *Non-smooth Thermomechanics: Physics and Astronomy* (Springer 2002)

C. Lexcellent, S. Leclercq, B. Gabry, G. Bourbon, The two way shape memory effect of shape memory alloys: an experimental study and a phenomenological model. Int. J. Plast. **16**, 1155–1168 (2000)

C. Lexcellent, *Les alliages à mémoire de forme* (Hermes- Lavoisier, 2013)

E. Patoor, A. Eberhadt, M. Berveiller, Potentiel pseudoélastique et plasticité de transformation martensitique dans les mono et polycristaux. Acta Metallurgica **35**(11), 2779–2789 (1987)

J. Perkins, R.O. Sponholz, Stress- induced martensitic transformation cycling and two-way shape memory training on cu-zn-al alloys. Metall. Trans. A **15A**, 313–321 (1984)

Chapter 7
Paul Ricoeur: "Memory, History, Forgetfulness"

Abstract In his book, "Memory, History, Forgetfulness" Paul Ricoeur is one of the first philosophers who tried to position memory in relation to the neuroscience discourse. He starts from the assumption that memory claims to be faith full to the past, although it relies on affection and sensitivity. Neuroscience addressed this issue under the title of mnesic traces which should not monopolize our attention. Phenomenologically, we do not know anything about the cortical-substrate of evocation.

Reading report by Pauline Seguin (2012)

Why Paul Ricoeur? Because he is surely one of the first philosophers who tried to position memory in relation to the neuroscience discourse.

"This new masterpiece by Paul Ricoeur (2000) is astonishing in many ways: Paul Ricoeur wrote it more than twenty-five years after his retirement from French university, and eleven years after his last class at the University of Chicago. In addition, this 660-page book is a formidable philosophical tour de force. He quotes and discusses the ideas of 213 authors" (Reagan 2008).

7.1 Introduction

Pauline Seguin tried to summarize the complex theories defended in the book "Memory, History, Forgetfulness" and to update the major lessons that can be retained to enlighten her study on the memory of decolonization in Indochina.

Let us briefly consider Paul Ricoeur:

Pauline Seguin (2012) wrote:

"Paul Ricoeur (1913–2005) was born on the eve of World War I, in 1913, and was a war prisoner in Pomerania between 1940 and 1945, during the Second World War. He thus immediately appeared as someone "profoundly marked by a tragic century that he lived through, was fully involved in its major issues, and aimed to clarify their meaning" (Ricoeur 2001). The dramas and tragedies of the twentieth century, to which he seemed to be very sensitive, profoundly marked his thought and his questioning, and this increased sensitivity was perceptible throughout his work.

© Springer Nature Switzerland AG 2019
C. Lexcellent, *Human Memory and Material Memory*,
https://doi.org/10.1007/978-3-319-99543-4_7

Influenced by the phenomenology of Husserl (1964) (of which he seemed to be a direct heir) and by existentialism, he constructed a philosophy of interpretation by integrating the contributions of psychoanalysis and Freud's thought, and appeared as one of the major representatives of contemporary hermeneutics. The great questions that landmarked his whole work were will, time, the subject, otherness, and values, in the perspective of an optimistic Christian humanism.

Time and history were one of his favorite research themes as early as the 1950s. He published "History and Truth" (Ricoeur 2001) in 1955, in which he attempted a delicate definition of the notion of truth in history, and delineated the outlines of objectivity in history (to be distinguished from objectivity in the exact sciences). In 1983, he continued his reflection by publishing "Time and Narrative", in three volumes (Ricoeur 1991), in which he wondered about the resources of narrativity, narrative, and fiction, and their impact on the narration of times gone by Ricoeur (2000).

Paul Ricoeur investigated the issues related to the relationships between the three gateways to the past: memory, history, and forgetfulness.

He did not seek to build a philosophy of history, or to establish a teleological reconstruction of history.

The book, published in 2000, may appear as the result of the sum of the reflections that the tragedies of the twentieth century and the questions of contemporary historiography inspired to the author, combined with the cross-reading of the various philosophers and thinkers who wrote on history. The author indeed strove to interweave the texts of many thinkers who wrote about memory, history, and its writing; as he said himself, he is an author "in front of whom all books are simultaneously open". He read everything from philosophers to sociologists, historians, linguists, and scientists. His study represents both the expression of its author's original theory and a summary of all that has been thought and published on the subject before.

To replace this book as far as ideas are concerned, it should be noted that regarding history, this study is in line with the works of historians of the late twentieth century (it refers in particular to Jacques Le Goff, Pierre Nora, Michel de Certeau) who questioned the practice of history and its writing. In philosophy, Paul Ricoeur's research echoed phenomenological reflections on the essence of memory phenomena, on the fundamental structures of historiography and the writing of history undertaken by philosophers such as Husserl or Bergson. In relation to social sciences, the author is in line with Foucault's and Halbwacks's heritage (Halbwacks is the author of works on issues of collective memory (Halbwachs 1950), and attempted to decipher the identity and political issues of uses and misappropriations, or abuse, of both memory and forgetfulness).

But Paul Ricoeur was above all a humanist and a militant philosopher who claimed his duty as a citizen. He wrote this book in reaction to Holocaust negationism, in the hope of being able to point out to his contemporaries the direction of a peaceful, reconciled memory and the direction of forgiveness.

He stated his commitment from the very first page. He specified that his book resulted from a reflection on the issues related to the links between memory and history, and was a response "to the troubles (aroused) by the disturbing show that supplied too much memory here, too much forgetfulness elsewhere, not to mention the influence of remembrance ceremonies and the abuse of memory and forgetfulness. The idea of a policy aiming for a fair memory was in this respect one of (his) overtly stated civic themes" (Ricoeur 2001).

This indignation in the face of "too much" seems to me quite important, as if through this permanent exaggeration, one wished to deny the substance of the subject or more prosaically to avoid the issue or the poison of memory!

Paul Ricoeur's book (Ricoeur 2000) raises the question of the representation of the past, 'what about the enigma of an image that presents itself as the presence of an absent item marked by the seal of earlier times? "He intends to give an account of this issue through a triple perspective: the phenomenology of memory, the epistemology of history, and the hermeneutic of the historical condition, where a reflection on the question of forgetfulness is inscribed."

The first question raises the whole issue of the present representation of something from the past. The object of the representation no longer exists, but the representation is in the present time.

"The phenomenology of memory proposed here is structured around two questions: what do we recall? who do we remember?" (Ricoeur 2000, p. 3)."

Within this heritage, he favored the request that any consciousness is the consciousness of something. "This objectal approach poses a specific issue regarding memory.

Memory is not basically reflexive, as the pronominal form "se souvenir" seems to indicate in French : does remembering something imply remembering oneself ?"

Ricoeur made a point of asking "what" before "who" despite philosophical tradition. The primacy given to the question "who" leads to a deadlock as soon as one takes into account collective memory. The question of who to attribute the action of remembering has to be left pending, and one must start with the question "what?". Remembering is to have a recollection or start a quest for a recollection. In this sense, the question "how" tends to follow from the question "what?". The question "who?" is focused on the appropriation of a recollection by a subject able to remember himself.

This was the path followed by Paul Ricoeur: from "what?" to "who?" via "how?" from the recollection of reflected memory via reminiscence.

Pauline Seguin slightly modified the question, which finally read "what do we remember when we remember?"

Paul Ricoeur meant to refer to the traditional philosophical debate about reality and fiction. It starts from the assumption that memory claims to be faithful to the past, although it relies on affection and sensitivity. When we remember, he wondered whether we remember this sensed impression or the real object from which it proceeds. Following Plato and Aristotle, Paul Ricoeur distinguished two types of memory: mn'em'e and anam'esis. Mn'em'e designates this sensitive memory which affects us unwillingly, whereas anamesis refers to what the author calls recall, which he means as an exercised memory, an active and voluntary search directed against forgetfulness. In fact, in this sense, one of the main functions of memory is to fight against forgetfulness, hence the idea of a duty of remembrance (Todorov 1995) referring in fact to a "duty not to forget".

7.2   Memory Abuse

Before dealing with memory abuse, Ricoeur considered the forms of successful memory, for example, memorization of a poem or of the grammar rules of a foreign language, or even the memorization technique (ars memoriae).

The main difference between memory and imagination, while both are related to the issue of the presence of something absent, is that memory is the guarantor of the past character of what it proclaims to remember. Memory is necessarily memory of something that is no longer, but has been, it therefore refers to true past events. But the imaging of the initial recollection supposes a reconstruction. This raises the question of the reliability of memory, and with it the question of its structural vulnerability. This vulnerability, which results from the relationship between the absence of the remembered thing and its presence as a representation, indeed subjects memory to multiple forms of abuse.

The author distinguished between three types of abuse: memory prevented, memory manipulated, memory compelled. By relying on the contributions of psychoanalytic theories, he meant by memory prevented the difficulty in remembering a trauma. Ideally, such a recollection requires the recourse to a remembrance process, via a mourning process, in order to be able to relinquish the lost object and tend towards a soothed memory and reconciliation with the past. At the scale of collective memory, traumatic memory may be a "wound of national self-esteem". If it is not the object of a remembrance process implying a real mourning process and placing events in a critical perspective, it exposes itself to the danger of what psychoanalysts call compulsion to repeat. This is particularly obvious in case of "too much memory" or repetition of funeral celebrations, as in the case of the Yasukuni Shrine in Japan. Only a mourning process and a critical perspective, based on remembrance attempts, will allow a society to tend toward a peaceful reconciliation with its past.

In the case of memory manipulated, the author refers to the ideological manipulations of memory. Power holders mobilize memory for ideological purposes "to serve the quest for, recovery of, or claim for identity." This type of ideological phenomenon aims to legitimize the authority of the power in place, to make it appear as a "legitimate power to be obeyed". The author posited the narrative style of the narrative as the main agent of the ideologization of memory. The narrative is by definition selection and coherence. Strategies of forgetfulness and remembrance therefore arise from the narrativity of the narrative. Official history is therefore also memory imposed, in that it is taught, "learned and publicly celebrated". Todorov and his works on memory abuses (Todorov 1995) pointed out that any work on the past is a work of selection and thoughtful combination of events with one another. This work is necessarily oriented, not towards a search for objective truth, but towards a search for the good (depending on the context, it may be the search for a certain social peace, the legitimation of the power in place, etc.). This thinker also revealed that manipulation of memory often tended to use victimization strategies, insofar as claiming to be a victim places the rest of the world in a position of indebtedness, and the victim's complaint, protest, or claim in turn appears to be legitimate. In this sense, manipulation of traumatic memory makes it possible to claim expectations from the future, because the memory of the traumatic past impacts the future.

The third form of abuse is memory compelled or commanded. This is what happens when official history is recited by schoolchildren or when national anthems are sung before sports competitions or at official commemorations, such as national holiday parades.

With memory compelled, the author addressed the question of the "duty of remembrance". In his lecture on fair memory, he was careful to make it clear that the duty of remembrance was not an abuse as such, but a real duty to do the victims and the cause (that brought about victims) justice, and to identify the victims and the perpetrators. Possible abuse grafts itself on the authenticity of this legitimate duty. In view of the historical conditions and the context in which this duty of remembrance is required, one is indeed able to grasp the moral stake, the meaning and the vision of the future that it carries. The idea of duty of remembrance necessarily involves the notion of debt, insofar as it places contemporaries in an indebted position towards those who preceded them. Ricoeur then assumed that there was a distinct but mutual building up of collective memory and individual memory, and he made the hypothesis of a triple attribution of memory, to ourselves, to close relationships, and to others (Ricoeur 2001).

In a recent book entitled "*Ma mémoire et les autres*" (Eustache et al. 2017) Francis Eustache asked the following question: "Does individual memory really exist?" The term individual memory or particular memory attributed to an individual retains a meaning (since each subject is unique), but results both from interactions with others and from the subject's truly personal, intimate story. In short, man is a social being. We will see later that Halbwachs (Halbwachs 1950) and Ricoeur (Ricoeur 2001) had largely answered the question.

But what is the relationship between individual memory and collective memory? How does one pass from personal recollections, which are characterized by their radical singularity, to the recollections of a whole community? After evoking St. Augustine's, John Locke's and Husserl's theories on individual memory, Ricoeur returned to collective memory with Husserl's idea of inter-subjectivity. An analogical transfer allows the transcription of the most important traits of individual memory, such as mine-ness, continuity, past-future polarity. This is why we can talk about common recollections re-staged in official rites, public holidays and permanent commemorations, such as cemeteries, monuments, and museums (Reagan 2008).

At that point, Ricoeur analyzed Maurice Halbwachs's theory (Halbwachs 1950) that memory is collective. In order to remember, we need other people. It is not an analogical passage from individual memory to collective memory. Any recollection is intertwined with other people's testimonies. There is an original ability for communities to preserve and recall common recollections.

Paul Ricoeur also wondered if there was an intermediary state between individual memory and collective memory. His answer was that close relationships, especially relatives, play this role. "Relatives, these people who count for us and for whom we count, are located on a range of variation of distances in the relationship between our own self and other people". They are the ones who remember our birth, which is the event that concomitantly starts our personal existence and our existence as a member of a community. Our birth certificate places us in time (date of birth), in space (place of birth), in a community, and above all in a family (Reagan 2008).

Ricoeur ended his phenomenology of memory with the following thesis: "It is not therefore with the sole hypothesis of the polarity between individual memory and collective memory that we must enter the field of history, but with that of a triple attribution of memory: to oneself, to relatives, to others."

Ricoeur started a very interesting discussion on Plato's pharmakon to determine whether writing is a cure or a poison for memory. To solve this problem, he sought to define the link between memory and archive or documentary evidence. He highlighted the link between inhabited space and historical time. Every recollection relates to a particular point in space (e.g., the house I once inhabited), and collective memory is always attached to a sacred or traditional place. This is why history is always closely linked to geography and cartography.

The origin of archive is testimony. "Testimony opens onto an epistemological trial that starts with declared memory, goes on with archive and documents, and ends with documentary evidence." Archive and the courts must both confront the issues of trust and suspicion. How can one verify a historical testimony? The same is true about memory, which depends on the witness's faithfulness and is subject to other people's corroboration. "Reliable witnesses are the ones who can maintain their testimony over time" (Ricoeur (2000), p. 206). Testimonies become archive as soon as they are in written form.

Archive is not only made of recollections and written testimonies; it is also a systematic classification of these documents with rules according to which they can be looked up. In addition, certain traces are not written, e.g., art objects, money, the remains of old buildings, tools, funerary objects. Today we can add (for example) fingerprints, photographic evidence, biological analyses such as DNA analysis to this list.

To pass from an archive or traces in general to documentary evidence, it is necessary to formulate a question, a hypothesis. For the historian, everything can serve as potential proof of evidence: parish registers, climate data, prices of agricultural products, daily life objects, court acts; we have all these documents in mind when we refer to archive. The historian must approach his field of study with a questionnaire, a hypothesis, a global idea of what he is looking for (Reagan 2008).

7.3 Memory, History, and Forgetfulness: All Interlaced

The analysis of the epistemology of historical knowledge reveals the problems arising from the relationship between practical knowledge of history and living experience. The author presupposed that there was competition and confrontation between history's intention for truth and memory's claim for faithfulness. The work of detachment, according to methods proper to historians, allows historical knowledge to become autonomous and to keep its distance from the experience of living memory. Much of the historian's material is made up of archives, themselves derived from testimonies by men of the past. This again refers to the question of the reliability of testimony, because it includes narrative and rhetorical components, and fulfills the requirement for consistency and the requirement to convince one's interlocutor. The historian must therefore go through a necessary confrontation between the different documents to be able to establish a probable and plausible account of events. The recourse to testimony is fully justified, since the subject of history is not the past nor time, but "men in time". Paul Ricoeur took care to clarify the difference between the historical fact and the actual remembered event, which can in particular cause conflicts between survivors' memory and written history.

The historical fact is constructed thanks to the work that brings it out from a series of documents. History says that such and such an event happened, "as it is told? That is the whole point." A historian cannot assert anything without evidence, but for a document to serve as evidence, the person who looks it up must ask it a question, and this question necessarily implies an underlying nascent explanation. Interpretation and its pitfalls are therefore present at all levels of the historiographical operation (i.e., at the documentary level, the explanatory level, and the level of the narrative representation of the past). Interpretation thus appears as a structural component of the intention for truth of all historiographical operations. However, the historian claims to "truly represent the past" (Ibid., 295); in this sense, history appears as the "critical extension" of the ambition of fidelity to the past times of memory, history is the "learned inheritor of memory". But Paul Ricoeur liked to question the ability of the historical discourse to truly represent the past, which refers directly to the dialectics of narration, and presumes a dual narrative-rhetorical component. Before being the object of historical knowledge, an event is first the object of a narrative (archive), hence the return of the aporia encountered on the issue of memory, namely that of the debate between reality and fiction. What is the difference between history and a story, if both of them narrate, i.e., tell a story? This is what the author called "the aporia of truth in history", which is particularly obvious when historians build different narratives from identical events. From there, the historian is not a neutral agent, he is a social being (Bergson 1896), in the position of a "committed spectator" (R. Aron). This makes borderline events, in which the historian clearly has a "responsibility towards the past", all the more obvious. The distinction between a historical narrative and a fictional narrative lies "in the nature of the implicit pact" made with the reader. It is indeed agreed that the historian deals with events, situations, and characters that existed before a narrative was built about them. Likewise, this pact also claims to correspond to and be in adequacy with the past. But this is an alleged, not necessarily effective fit between historical representation and the reality of the past.

In the section about the "hermeneutic of the historical condition", the author highlights the interdependence between the historical reading of the past, the way in which the "present time is lived and acted" and the expectations from the future. "The projection of the future is (actually) [...] tightly linked to the retrospection of past times'. Yet a historian, like a judge, occupies the position of a third party, and in fact aspires to impartiality. But this is a necessarily unfulfilled aspiration, in that total impartiality is impossible. The historian cannot and will not make a historical judgment (and even if he did so, historical judgments are in essence provisional and controversial). In the same vein, just as it is impossible to reach absolute impartiality, the historian does not have the means to write a global history that would erase differences between points of view, a unique history that would embrace the history of performers, the history of victims, and the history of witnesses (Ibid., 334). That way Paul Ricoeur raised the question of the historiographical treatment of the unacceptable: "How is it possible to deal with the extraordinary with ordinary means of historical understanding?" These types of borderline events, whose author was both a witness and a victim, make it all the more apparent that the intervention of subjectivity in history is unavoidable. Interpretation is present throughout historiographical operations, and it is always possible to interpret an event differently. Controversy seems inevitable, and history is bound to perpetual revisionism (ibid., pp. 447–448).

7.4 Forgetfulness

Forgetfulness is related to the issues of memory and fidelity to the past (Ricoeur 2000). It encompasses the issue of forgiveness, in that forgiveness appears as the last step of the forgetfulness process. Forgiveness is part of the issue of guilt and reconciliation with the past, but both of them tend towards the aim of a soothed memory. In its current meaning, forgetfulness is first of all felt negatively, as an attack on the claimed reliability of memory. Yet, according to the author, two kinds of forgetfulness should be distinguished. On the one hand, the negative side of forgetfulness is a source of anxiety, it is oblivion that "erases traces". On the other hand, the positive side of forgetfulness is "in-store forgetfulness", a source of pleasure, when, like Marcel Proust with the famous madeleine of "In Search of Lost Time", we remember what we have seen, heard, tested, acquired. This idea echoes the theory of a reversible forgetfulness, supported by Bergson in "Matter and Memory" (Bergson 1896), or refers to the hypothesis of the unconscious and to the idea of the unforgettable, represented by Freud. The memory work is directed against oblivion through erasure of traces. Forgetfulness is therefore linked to memory, it is in a way its dark side, maybe even the very condition for memory to exist and be used. "Recollections are only possible on the basis of forgetfulness, not the other way around". And, it is as a negative counterpart of memory that forgetfulness can be the object of the same abuses as memory.

We should note that forgetfulness is equivalent to amnesia of certain metals that lose their memory or training if they are left unused in a closet for a long time.

This was the first time in my readings that I found forgetfulness and forgiveness correlated.

In the case of the prevented memory of a traumatic event, the compulsion to repeat equals to forgetfulness in that it prevents the awareness of the traumatic event. Regarding manipulated memory, the abuse of memory is also an abuse of forgetfulness (for memory is a narrative, so it is by definition selective). It is always possible to tell things differently, "by deleting parts, and by shifting stresses". But for the author, this "insufficient memory", if it is imposed from above, is comparable to a kind of "semi passive" forgetfulness, insofar as it implies a certain complicity of social actors who demonstrate a "will-not-to-know".

In the context of controlled and institutionalized forgetfulness, the author attempted to deal mainly with the case of amnesty, whose phonetic proximity with amnesia raised his questioning. Amnesty is a form of "institutional forgetfulness" for him, a "denial of memory [...] (that) in truth moves away from forgiveness after proposing to simulate it". Amnesty boils down to acting as if nothing had happened, it is an injunction of the State to "not forget to forget". But it turns out that the price to pay in case of institutionalized amnesia is high, because collective memory is deprived of the salutary identity crisis that would allow society to lucidly reclaim for itself the past and its traumatic burden through processes of memory and mourning both guided by the spirit of forgiveness. Forgetfulness, according to Paul Ricoeur, has a legitimate and salutary function, not as an injunction, but as a vow. If there is a duty of forgetfulness, it is not "a duty to silence evil, but to voice it in a soothed way, without anger."

7.5   A Profound Text

While underlining that Paul Ricoeur had written a very profound text (Ricoeur 2000), Pauline Seguin (Seguin 2012) criticized two aspects of his work, i.e., an "ambition of truth in history" and a claim to "fidelity of memory" without questioning it. Yet, the notion of truth appears to be highly questionable because it is fundamentally relative. The historian is indeed willing to prove what he believes to be the truth, relying on true documents inherited from the past, but does that imply that the theory he supports is true, absolutely true? If, as we often hear, history is the story of winners, the original ambition of historians may well have been less the truth than the demonstration of the greatness and legitimacy of their backers.

Truth and history
1. Agreeing about words. The word truth is meant here as material truth, i.e., what Jules Lachelier called "agreement of thought with thing", and Krzysztof Pomian "the truth as adequacy of knowledge to reality". From the historian's point of view, what

is considered as true is what corresponds as much as possible to existing or past reality. This reference to an "objective" reality is consubstantial to history:

- Langlois and Seignobos (Introduction to Historical Studies, 1898) defined history as "a representation of a past reality", and while asserting that history is necessarily subjective, stated that "subjective is not synonymous with unreal"
- Rejecting "negationism", a group of historians from Lyon (Le Monde, 29/04/1993) affirmed that there exist facts that cannot be reduced to any historical subjectivity"
- Paul Ricoeur (Time and Narrative, vol. 1, 1983) (Ricoeur 1991) wrote: "Even if the past is no more and if, according to Augustine's words, it can be reached only in the present of the past, that is to say through the traces of the past that have become documents for the historian, nevertheless the past did take place".
- Roger Chartier (At the Edge of the Cliff, 1998): "This reference to a reality located outside and before the historical text, whose function is to restore it in its own way, has not been abdicated by any of the forms of historical knowledge. Better still, it is what makes up history as opposed to fable and fiction."

The word History is polysemic in French, as it designates both past reality and the knowledge built up from it. But these two "stories" are literally anachronistic, delayed : history is a writing of the past built up in the historian's present. It is in this time gap that the issue of the truth of history essentially lies. And this time gap is one of the reasons why the scientificity of history is problematic.

We shall begin with a "counter-relief" approach to the question of the truth of history by localizing it in relation to the scientific process and to literary creation.

At this point, we have to focus on Bruno Bachimont's issue (Bachimont 2010) which is to determine how we can fit our relationship with the past with this archived memory. The issue is also to understand what the digital world is doing to archive, how it modifies our notion of memory.

History: a science?
Two objections are often expressed: except in the case of "present time" or "immediate" history, the historian cannot directly observe his object: history is knowledge through traces. In history, one cannot experiment, reproduce phenomena (as in laboratory), change the parameters (in spite of the attempts of the American "new economic history").

A brief chronology of history/science relationships—At the end of the 19th century a brief "scientist" phase was illustrated, among others, by Numa Fustel de Coulanges in the preface of "La Monarchie franque" 1888)

"Facts to be analyzed, brought closer, linked together [...]. The historian has no other ambition than to see facts properly and understand them accurately [...] He seeks and attains them through careful observation of texts, as the chemist finds his own facts in meticulously conducted experiments."

This claim to be a "pure science" was quickly denounced by Langlois and Seignobos (op.cit.), who stated that "Science is an objective knowledge based on real analyses, real summaries, and real comparisons; the direct sight of objects guides the scientist and dictates him the questions to be addressed. [...] In history one sees nothing real other that written paper..." Historical analysis "is no more real than the sight of historical facts; it is only an abstract process." They still write that "facts that we have not seen, described in terms that do not allow us to represent them exactly, those are historical data". And they concluded: "By the very nature of its materials, history is necessarily a subjective science. It would be illegitimate to apply the rules of real analysis of real objects to this intellectual analysis of subjective impressions. History must therefore refrain from the temptation to imitate biological sciences. History is not an art, it is a pure science [...]; it consists, like any other science, in taking notes."

The same old tune is on the air again: if some show that history is a "science", it will sound more serious and therefore more credible!

As regards Aristotle's claim (Aristote 1866) of fidelity to memory, this is an equally questionable point. Memory seems on the contrary to assume its subjectivity and to claim its sensitivity. It certainly knows itself to be more malleable and questionable than what Paul Ricoeur seems to want to say. Adjustable at will according to external requirements, it could be believed to have difficulties in disapproving Nietzsche's maxim, "I did that", my memory says. "Impossible!" my pride says, and pride persists. In the end, it is memory that gives way (Nietzsche 1990).

There is even a national association for loyalty to the memory of General De Gaulle. We can assume fidelity to the values of gaullism and the general's actions during the Second World War such as the Call of June 18,1940, but we should mind that it does not turn unalterable as time passes!

Customer loyalty includes all the actions implemented by a company to ensure that its customers remain loyal to it, and keep consuming its products or services. The purpose is to be able to establish a long-lasting relationship with each of the customers. If we replace the word "client" by the word "memory", things grow trickier, but motives can be the same within an association. Customer loyalty appears to have erased loyalty to memory.

7.6   Memory and Imagination

We indistinctly say that we represent ourselves a past event or have an image of it. A long philosophical tradition has turned memory into a province of imagination already treated suspiciously long before, as shown in Montaigne and Pascal. However, imagination per se is located at the bottom of the scale of the modes of knowledge. And since memory, which is also considered as a mode of education because of the memorization of traditional texts, has a bad reputation: "Nothing comes to the rescue of memory as a specific function of access to the past" (Descartes, "The Speech of the Method" 1637), imagination and memory have to be decoupled. The guiding idea is the difference between them: the function of imagination is directed towards the fantastic, the fiction, the unreal, the possible, the utopian, while the function of memory is directed towards a previous reality, an anteriority making up a temporal mark par excellence of the "remembered thing", of "the remembered" as such (Ricoeur 2000, p. 6).

We must turn to Aristotle for a specificity of the properly temporalizing function of memory to be acknowledged: "memory belongs to the past" (Aristote 1866).

The permanent threat of a confusion between remembrance and imagination, resulting from the transformation into memory images, affects the ambition of fidelity on which the true function of memory is based (Ricoeur 2000, p. 7).

7.7   A Phenomenological Sketch of Memory

Two observations can first be made:

The first observation is that many authors tend to apprehend memory from its deficiencies, and even its dysfunctionings (as very often in neuro-science where brain deficiencies are measured, for example).

Nobody can blame imagination for a similar reason insofar as its paradigm is made of the unreal, the fictitious, the possible.

"To put it bluntly, we have nothing better than memory to mean that something happened, took place, occurred before we declared that we remembered about it." (Ricoeur 2000) (ibid, p. 26).

As Ricoeur showed, testimony is the fundamental transitional structure between memory and history.

The second observation is an assertion by Aristotle, according to which "memory belongs to the past". But belonging to the past can be expressed in many ways.

"We speak about memory and recollections." Radically speaking, it is a phenomenology of memory that is addressed by Ricoeur.

"Memory is in the singular, as a capacity and effectuation, recollections (memories) are plural: we have a certain number of recollections."

The first pair of opposites is composed of the habit/memory pair (See Bergson: "Matter and Memory" Bergson 1896).

The second pair of opposites is composed of the evocation/search pair.

Evocation is defined as the current occurrence of a memory. Therefore, as opposed to search evocation is an affection.

Evocation carries the load of enigma, namely the presence right now of the absence previously perceived, tested, learned.

This enigma must be temporarily dissociated from the question posed by the persistence of the first affection. This persistence is illustrated by the famous metaphor of the seal imprint and consequently the question of whether the accuracy of a recollection consists in the eikon (image, representation) resembling the first imprint.

More or less the same question arises when one educates a material through repeated or cyclic mechanical loading.

"Neuroscience addressed this issue under the title of mnesic traces, which should not monopolize our attention; phenomenologically, we do not know anything about the bodily- more precisely the cortical - substrate of evocation; nor do we have a clear overview of the epistemological regime of the correlation between the formation and the activation of these mnemic traces on the one hand and the phenomena which fall under the phenomenological view on the other hand" (ibid, p. 33)

The situation has not made any progress in the science of materials either.

Forgetfulness can be designated as that against which the recall effort is directed.

"We seek what we feared to have forgotten temporarily or forever, without being able to decide, on the basis of the ordinary experience of recalling, between two hypotheses concerning the origin of forgetfulness: is it a definitive erasure of the traces of previous knowledge, or of a temporary, possibly surmountable impediment that opposes their resuscitation?" (Ibid, p. 33).

The main distinction is between laborious recall and instantaneous recall (Bergson, "Spiritual Energy", pp. 932–938 (Bergson 2012)).

Instantaneous recall is the degree zero of the search, in Bergson's fight against reduction.

If the question of recall comes ahead of the examination of the various kinds of intellectual works, it is in the form of a gradation "from the easiest, which is reproduction, to the trickiest, which is production or invention."

This refers to experimental observations on a trained shape memory alloy (SMA): the easier the training of the alloy, the faster it loses its memory; the more difficult training is, the longer it keeps its memory!

Special emphasis should be laid on the distinction introduced by Husserl in "Lessons for a Phenomenology of the Intimate Consciousness of Time" (Husserl 1964) between retention (primary recollections) and reproduction (secondary recollections).

The radical question is "the origin of time", i.e., what is temporal duration? In sum, qualifying as originary the moment of the past when retention occurred equals to denying that retention is a representation through images.

Casey's book (Casey 1987) is a plea for what Ricoeur calls "happy" memory, as opposed to descriptions motivated by suspicion or by the excessive primacy given to the phenomena of memory failure or even memory disorders.

References

Aristote, *De la mémoire et de la réminiscence.* (Ladrange, Paris 1866). Traduction de Jules Barthélemy-Saint-Hilaire, -55 avant JC

B. Bachimont, La présence de l'archive: réinventer et justifier. La revue de l'Association pour la Recherche sur les sciences et de la Cognition **53–54**, 281–309 (2010)

H. Bergson, *L'Energie spirituelle* (Payot, 2012)

H. Bergson, *Matière et Mémoire. Essai sur la relation du corps à l'esprit* (1896)

E.S. Casey, *Remenbering. A Phenenological Study* (Indiana University Press, 1987)

F. Eustache, H. Amieva, C. Thomas-Anterion, J.G. Ganascia, R. Jaffard, C. Peschanski, Stiagler. B., *Ma mémoire et les autres.* ESSAI LE POMMIER! (2017)

M. Halbwachs, *La mémoire collective* (PUF, Paris, 1950)

E. Husserl, *Leçons pour une phénoménologie de la conscience intime du temps (trad. fr. de H. Dussort).* PUF coll "Epiméthée" (1964)

F. Nietzsche, *Maximes et intermèdes 4e partie : "Par delà le bien et le mal"* (Edition Christian Bourgeois, Paris, 1990)

C. Reagan, Réflexions sur l'ouvrage de Paul Ricoeur: la mémoire, l'histoire, l'oubli. Transversalités **106**, 165–176 (2008)

P. Ricoeur, *La mémoire, l'histoire, l'oubli* (Editions du Seuil, Points Seuil, 2000)

P. Ricoeur, *Temps et Récit* (Editions du Seuil, Paris, 1991)

P. Ricoeur, *Histoire et Vérité* (Editions du Seuil, Paris, 2001)

P. Seguin, C.r. de lecture du livre de Paul Ricoeur "la mémoire, l'histoire, l'oubli" Paris editions du seuil 2000. In Mémoires d'Indochine (2012)

T. Todorov, *Les abus de la mémoire* (Arléa, Paris, 1995)

Chapter 8
Memory and Forgetfulness: From Psychoanalysis to Neuroscience

Abstract The interweaving of forgetfullness with memory explains the silence of neuroscience on disturbing and ambivalent experience of ordinary forget-fullness. Eve Suzanne poses a question: can neurobiology explain the functioning of psychic life ? In short, we can admit that we are not determined by our neurons.

"I am far from thinking that the psychological floats in the air and has no organic foundations", Zygmund Freud (September 22, 1898).

It can be useful today to attempt to shed light on the decisive arguments of psychoanalytic experience using anti-scientific data. The question of memory occupies a central place in the two fields we are trying to cross-study. The memory of compulsion to repeat and the memory of recollection of the analytic cure go against the implicit and explicit memories of the neurosciences, with forgetfulness as a corollary Corcos (2008).

In remembrance, repetition, perlaboration (1994) , Freud stated: "It is in the handling of transfer that we find the main means of stopping the compulsion to repeat and transforming it into a reason to remember." The present objective is to read this sentence again in the light of current neuroscientific knowledge about memory.

8.1 Different Forms of Memory

Neuropsychology distinguishes between several forms of memory, first of all short-term memory and long-term memory. We will not dwell on short-term memory that does not belong to our subject, but we will address long-term memory, the memory that psychoanalysis calls on.

This long-term memory is itself subdivided into two more types:

- explicit memory (also called declarative memory),
- implicit memory (also called non-declarative memory).

© Springer Nature Switzerland AG 2019
C. Lexcellent, *Human Memory and Material Memory*,
https://doi.org/10.1007/978-3-319-99543-4_8

Explicit memory refers to conscious traces of past experiences. It involves conscious recollection and remembrance. It includes two more memory types called episodic memory and semantic memory.

Episodic memory is the autobiographical memory, the memory of events, the things lived by the subject (for example, the night of my birthday party in year XXXX). Semantic memory deals with more general, more abstract contents related to knowledge, culture (for example, what I know about such or such subject or the vocabulary I have in store).

Implicit memory refers to the unconscious effects of past experiences. It is the memory we are first interested in. It is described here in its two dimensions: the procedural dimension (memory of motor skills), and the emotional dimension.

To illustrate the definition of implicit memory, we can report the experience of B. S., a famous patient of the Swiss doctor and psychologist Edouard Claparède (who ensured the dissemination of Freud's work in French-speaking Switzerland at the beginning of the twentieth century). In the early 1900s, Dr. Claparède followed the progress of B. S., a patient with Korsakoff syndrome, which results in severe amnesia. B. S. never recognized Dr. Claparède, who had seen her for many years. During a consultation, E. Claparède hid a pin in his hand as he greeted B. S., which caused her to start. The next time, when Dr. Claparède reached his hand out to his patient, she immediately withdrew hers. When he asked her the reason for this reaction, she could not explain why. Traces of the past experience had thus been inscribed in her brain circuits, without any conscious recollection.

This anecdote found further developments in the investigations and the elaboration of the concept of implicit memory. The impact of all so-called subliminal perceptions is based on implicit memory. Numerous experiments showed that in the absence of conscious perception, subjects were able to reproduce the content of a stimulus that they considered they had not perceived. As described above, implicit memory is subdivided into procedural memory (motor skills) and emotional memory. It is its emotional quality that matters to us here. Its nerve center is the amygdala, which receives information through two pathways: a short thalamus-amygdala pathway and a long cortical pathway.

8.2   The Release of the Compulsion to Repeat, and Forgetfulness Through Retroactive Interference

Forgetfulness takes different forms in neuropsychology. One of them is forgetfulness through retroactive interference. It consists in the fact that a new datum tends to erase and take the place of an old mnemic trace belonging to the same field of information. How is it possible to match this definition with the release of the compulsion to repeat into transfer? Repetition is expressed everywhere, in all places of the subject's life, but it is during the therapy that it unfolds in the most intelligible way. Transfer is based on a postponement, a substitution, and it is within the framework of the transferential relationship that the compulsion to repeat will act. "We allow it access to transfer, a kind of arena where it will be allowed to manifest itself in almost total freedom and where we ask it to reveal to us all the hidden pathogenic aspects in the subject's psyche" Freud (1994).

The compulsion to repeat is brought back into play on the analytical ground, and the analyst presages a response to this repetition. The response is expressed within the framework he sets, his attitudes and his interpretations. The analyst thus embodies the response to repetition, and the analytic therapy makes it possible to replay the game. The transferential relationship welcomes a new publication of the identical, and through its unprecedented response, it proposes to develop an elsewhere, another thing, a new path that takes the place of this redundant memory.

It appears to us that this new path is comparable to the one that acts in retroactive interference. Hence, this lovely paradox: psychoanalysis allows us to forget. It makes it possible to forget certain traces of implicit memory by inscribing new ones, to erase certain traces of this unconscious memory that bring constraint and suffering. Psychoanalysis inscribes these new traces by means of everything that is intended for the patient but not made of interpretations stricto sensu. We will not mention all the technical nuances that distinguish between intervention, construction, interpretation... We will only evoke what expresses the analyst's implication in the analysis, without upsetting the transferential movement or bypassing the patient's speech.

We address what is received as (a) nothing by the patient in analytical work, or going with all the rest, i.e., what is not perceived as meaningful. In our opinion, these "corrective experiments", as Godfrind (1994) called them, permit forgetfulness through retroactive interference. However, we prefer the term "analytical surplus value" to "corrective experiments". This analytical surplus value is of course received by what makes up the most ancient and most profound traces of the patient's self. What is received in these terms is given throughout the analysis via multiple verbal and infra-verbal afferences, which contribute to the patient feeling safer in his/her body and in his/her psyche:

the constancy and safety of the framework,
the analyst's voice and prosody,
the fact that he/she remains benevolent,
the fact that he/she keeps his/her place,
the fact that he/she shows that he/she understands,
the fact that he/she listens,
the fact that he/she keeps silent,
the fact that he/she feels like talking...

We should recall that this securing only has a mutative effect in transfer, and in response to the patient's repetition. It differentiates it from the agreed securing of other psychotherapeutic approaches. We will not go back to Winnicott's contribution Winnicott (2000) to the question of internal safety, but we will be satisfied with quoting him "If a patient needs quiet, then nothing can be done except giving it to him/her. If we do not meet this need, anger does not ensue, but one simply reproduces the situation of environmental deficiency that stopped the growth processes of the Self." (Winnicott 1954).

This securing offered to our patients is reprinted on top of the painful emotional experience and allows its erasure through forgetfulness. Lived events are transformed, pain subsides, but as for history, it is not subjected to forgetfulness. On the contrary, the novel builds up, and so does memory. For another property of the analytical experience is to transform the lived events of implicit memory into an autobiographical construction of episodic memory.

8.3   Transformation of Repetition into a Recollection

In neuropsychology, episodic memory is the memory of events. It is not considered as an identical reproduction of the real. Neuropsychology far from ignores the self-constructed dimension of memory. Thus, while implicit memory emerges from the effects of episodic memory in terms of compulsion to repeat, episodic (autobiographical) explicit memory is enriched. The implicit, unconscious and acted memory of the compulsion to repeat is transformed into a conscious remembrance of autobiographical memory.

Neurosciences, like psychoanalysis, recognize the possible connections between these different forms of memory. This transcription of the compulsion to repeat, that is, its transformation into an autobiographical memory, is each time a symbolizing progress. This autobiographical sense-making is recognized by some neuroscientists as a vital need for human consciousness processes. Lionel Naccache (2006) tells the story of a patient, Mrs. R. M. B., suffering from neurological disorders following aneurysm rupture. The brain lesions in question (on the inside of the left frontal lobe), in addition to the disturbances of memory and attention that they produced, profoundly modified the emotional and affective character of this patient (a classic neurological consequence of this type of lesion). She then "exhibited a behavior of deep emotional indifference" and lost interest in the fate of her children although she had been an attentive mother until then. When she was asked about it, she would construct an interpretation of her troubles, saying that having been so close to death, her conception of things in life had changed.

For this neurologist, this patient, among other examples he cited, proved the inevitable requirement of fiction from the conscious process. This need is not conceived as exclusive to neurological patients, but is attributed to the ordinary needs of conscious neuro-cognitive functioning. Autobiographical episodic memory, and thus construction in analysis, respond to the perceptual contribution of reality, nourished by this fictional necessity.

Psychoanalysis works on a form of metabolization of memory, capable, as some neuroscientists say, of "interpreting" and "giving meaning". As we just saw, neuroscientific and psychoanalytic terminologies sometimes even overlap, just as the forms and convolutions of memory intersect with the lines of force of psychoanalysis.

8.4 Summary

Corcos (2008) proposed a few possible crossroads between psychoanalysis and neuroscience in the field of memory.

They concern the possible analogy between the compulsion-to-repeat concept and implicit memory in neurosciences, the similarity between forgetfulness by retroactive interference in neuropsychology and the erasure of traumatic effects through transfer, and the process of transforming repetition into memory, comparable to the transformation of implicit memory into autobiographical memory.

This came as a surprise to me because the neuroscience approach, which is more technical with a scientific ambition, is often opposed to the psychoanalytic approach, which is more human.

8.5 Forgetfulness and Erasure of Traces

It is customary in neural sciences to directly tackle the problem of mnemic traces with a view to localizing them, or subordinating the questions of topography to those of connectivity and hierarchy of synaptic architectures. Thus, one passes to the relationships between organization and function, one identifies the mental equivalent of the cortex in terms of representations and images, among which mnemic images. Forgetfulness is then in the neighborhood of dysfunctionings of mnemic operations, on the borderline between the normal and the pathological (Ricoeur 2000) (ibid., p. 543).

For Ricoeur, this program is scientifically irreproachable. But the philosopher's questions are in a different vein. For example, one wonders what makes a mnemic trace recognizable as such other than the relationship to time and the past. But for the phenomenologist, this relationship is specified by the central issue of the imaged recollection. As we previously explained, the philosopher's role is to link the science of amnesic traces with the issue of the representation of the past.

A question arises regarding forgetfulness: what kind of dysfunctioning is it? Could it be a similar dysfunctioning to amnesia cases belonging to the field of clinical cases? Ricoeur tried to hold on to the semantics of the discourses held by neuronal sciences on the one hand, and by philosophers claiming the triple heritage of French reflective philosophy, phenomenology, and hermeneutics on the other hand. The object-body (known in nature sciences) is semantically opposed to the lived body, the proper body, my body (from which I speak). Only at the end of a long route is there "the" brain, the object of neurosciences. Neurosciences take for granted the process of objectification, which remains a considerable, in many respects badly solved problem for hermeneutical phenomenology (ibid., p. 545). The scientist may allow himself to say that "man thinks with his brain"; for the philosopher, there is no parallel between the two following sentences: "I take with my hands", "I understand with my brain."

The scientist still respects the limits of this causal discourse when he confines himself to talking about the "contribution" of such or such cortical area, the implication or even the "responsibility" of such or such neural circuit, or declares that the brain is "concerned" by the appearance of such or such psychic phenomena.

Once on his own ground, the neuroscience man claims a less negative use of the causality between structure or organization and function.

Thus, a step can easily be taken to relate cortical traces to cultural traces. This is how the neurologist authorizes himself to consider the brain as the seat of images despite the reservations from the philosopher's semantic rigorism.

Neuroscience, one is tempted to say, does not in the least directly contribute to the conduct of life. This is why we can develop an ethical and political discourse on memory and conduct cutting-edge scientific activities in many human sciences without even mentioning the brain.

It is striking that the works directly devoted to memory and its distortions devote much effort to what Buser (1998) called a taxonomy of memory or rather memories; how many memory systems are required, should we count them? As a reminder, taxonomy is the science of the laws of classification.

The very meaning of the notion of trace in relation to the past has to be clarified.

Shall we pass from the metaphor of the wax imprint to that of the graphism of paintings?

"What causes the inscription to be both present as such and to be a sign of the absent, the previous?"

Shall we invoke the "stability of traces" in the same manner as hieroglyphs?

The very same question arises concerning shape memory alloys: when we educate an SMA, we "build" a microstructure of defects that carries the memory. These defects can be qualified as mnemic traces. When the material has "lost its memory", the traces in the microstructure have disappeared. This can be interpreted as forgetfulness. In order to shed light on this, we examined an example showing the effects of a thermomechanical treatment on the internal structure of the material (Perkins and Sponholz 1984).

But we still need to speak about forgetfulness! Clinical studies only address the precise subject of forgetfulness as related to dysfunctionings or "distortions of memory" (ibid., p 552). But is forgetfulness a dysfunctioning, a distortion? When we leave an SMA in a closet for a long time and it loses its memory, what kind of forgetfulness is involved? It can be definitive oblivion assignable to erasure traces.

In short, forgetfulness is often lamented in the same way as aging or death: it is one of the figures of the inevitable, the irremediable. Let us note here that even materials age and more or less no longer fulfill the functions originally assigned to them!

"The interweaving of forgetfulness with memory explains the silence of neuroscience on the disturbing and ambivalent experience of ordinary forgetfulness. But the first silence here is that of the organs themselves. In this respect, ordinary forgetfulness follows the fate of happy memory: it is silent on its neuronal basis (p. 553)."

8.6   Forgetfulness and Persistence of Traces

As said earlier, the notion of trace is neither reduced to documentary traces nor to cortical traces; both types consist of external marks: that of the social institution for archive, that of biological organization for the brain. There remains a third kind of inscription, i.e., the persistence of the first impressions as passivities: an event struck us, moved us, affected us, and the affective mark remains in our mind. There is no contradiction between the statement about the ability of inscriptions-affections to remain and last on the one hand, and knowledge about cortical traces on the other hand (Ricoeur 2000 p. 554).

The third kinds of inscription can be described as an affective or psychic trace.

"On the one hand, I trust the "bodily machine" in the operation of happy memory; but I am wary of its poorly controlled resources of nuisance, anxiety and suffering. On the other hand, I trust the original ability of affections-inscriptions to last and remain; without it I would have no access to the partial understanding of what the presence of absence, anterior distance, and temporal depth mean" (Ibid, p. 555).

Which life experiences can be considered as confirming the hypothesis of the survival of impressions-affections beyond their occurrence? The original experience in this respect is that of gratitude, that little miracle of happy memory. This recognition can take different forms; it already occurs during perception: a being was present once, then he/she was absent, then he/she came back. In many ways, knowledge is recognition.

8.7   Neurosciences, a Reductionist Position? By Eve Suzanne

8.7.1   Dualism and Neurobiological Reductionism

According to neurosciences, all of our mental states come down to the activity of our neurons, that is to say, to chemical reactions. The dualist theory that poses the soul, i.e., consciousness, as immaterial and therefore not reducible to mechanisms such as those to which the body is subjected, is clearly rejected. On the contrary, psychological facts are subjected to the same laws as neuronal facts. Besides, we can speak of eliminative reductionism: psychology will be completely replaced by neurobiology once neurobiology has progressed far enough on the path of knowledge (Suzanne 2009).

Reductionism, as defended by neurobiology, consists in placing consciousness and neurons at the same level. In other words, human psychic life is fully understood based on the neuronal chemical processes that occur in the human brain.

Thus, authors like Jacques Monod or Francois Jacob in the line of La Mettrie developed an extreme reductionist position based on neuroscience and its spectacular advances.

Thus, according to them our will, our desires, our convictions, all that accompanies and inspires our acts systematically depends on the organization of our proteins.

Jacob added to this subject and wrote that "between the living world and the inanimate world there is a difference, not in nature, but in complexity". The living is a machine. Monod conceived it as a mechanical machine: for him, the living was a machine controlled by its DNA and the cell was comparable to a small chemical plant. As for Jacob, he rather saw the living as a cybernetic machine (the brain was like a computer, which later supported the theory of artificial intelligence).

Descartes also thought that the difference between the body and the machine was not in nature, but only in the degree of complexity.

The philosopher J. R. Searle wrote as follows: "The whole of our conscious life is determined by these elementary processes." By "these processes", the author referred to brain processes, i.e., the chemical reactions that occur at the level of neurons and synapses. Knowing how these neurobiological processes cause consciousness seems unsurpassable.

This is why some great scientists remained dualists, like Sir John Eccles, a neurobiologist who received a Nobel Prize in 1963 for his work on synapses. He wrote "the brain is a machine that a ghost can operate" (in "The Wonder of Being Human: Our Brain and Our Mind", 1984). In fact, the brain was supposed to be on one side and the soul on the other side: God was believed to intervene three weeks after fertilization to infuse a soul into the fetus and connect it to the brain.

In this regard, we can say that we are in cloud cuckoo land!

Yet Searle opposed any form of dualism in the understanding of the psychic and biological life of Man. One should not consider that there are cerebral processes (the cause) on the one hand and consciousness (the effect) on the other hand. These are not two distinct events: these neuronal processes are not an entity standing apart but form "a feature of my brain in the present moment". In the case of consciousness, it is considered to be one of the functions of the brain in the same way as respiration or digestion.

8.7.2   Changeux and the Concept of Brain Plasticity

Paul Ricoeur considered that Jean-Pierre Changeux, his interlocutor during their debate on the question of the implications of neuroscientific advances in the ethical field. Changeux and Ricoeur (1998) escaped a merely reductionist position. Indeed, Jean-Pierre Changeux argued that he did not adhere to the kind of reductionism supported by part of the profession. Since the days of Monod and Jacob in the 1970s, neuroscience has made spectacular progress and is still growing exponentially. Now, these advances allowed Changeux to develop a third approach of reflection that psychoanalysts also considered with the emergence of the concept of brain plasticity.

Research indeed showed that the brain of certain patients could be reorganized if parts of it were no longer functional following accidents or diseases.

Changeux gave the example of a child who was born incapable of speaking because the area where language was to develop was malformed. But once placed in a stimulating environment, his brain was able to get reorganized so that the child was able to develop speech. These observations revive an old debate between nature and nurture. Two opposite opinions exist in that field: some insist on considering that our brain is fully determined by genetics, while others refuse this determinism and consider the possibility for our brain to build up according to the hazards of life, that is to say in a random way.

Changeux considered that both theories were right and wrong at the same time. He considered that there are invariants in our brains (like vision) that remain and persist in all of us, and that at the same time there are processes of plasticity at the level of molecules, synapses, networks of neurons, in short at different scales. "The brain is not a rigidly wired automaton. On the contrary, due to its plasticity, each brain is unique."

Brain plasticity makes it possible to think of each individual's singularity, which is determined in relation to their brain, but the brain itself changes according to the environment in which individuals live and to the experiences they go through.

8.8   The Case of Psychoanalysis, by Eve Suzanne

Can neurobiology perfectly explain the functioning of psychic life?

The underlying difficulty is to determine which of psychoanalysis or neurobiology is best able to provide such an explanation.

However, perhaps this opposition was not obvious at first, and perhaps it still is not obvious: Sigmund Freud was convinced that one day biology would finally validate the assumptions on which psychoanalysis and especially the unconscious are based.

8.8.1   Dualism and the Unconscious

Andre Green, a former Lacanian (from 1955 to 1967) psychoanalyst was particularly virulent towards certain claims of neuroscience. He was opposed to a fundamental aspect of their approach which placed brain activity and psychic life at the same level, and stated that understanding brain activity automatically allowed for understanding of psychic life. In an article written in 1992 (Green 1992), the author spoke of "a frenzied denial of the complexity of psychic functioning and in turn of the unconscious [...], by the defenders of the cause of the brain, namely neurobiologists, psychiatrists, and neurologists". For neurobiologists, all psychological disorders have an exclusively organic cause, otherwise the disease is imaginary.

Against all odds, neuroscience itself restored the primacy of the unconscious thanks to the cognitive unconscious. But far from rehabilitating the Freudian unconscious, the cognitive unconscious was aimed at definitively sending the Freudian unconscious into oblivion. Indeed, in the cognitive unconscious, there is no repressed desire, no interpretation of our dreams... On the contrary, it falls into the same reading grids as consciousness, as one more characteristic of our brain activity. The majority of our neural processes can be described as unconscious.

The answer to the question of whether Freud can be integrated into neuroscience is no, simply because psychoanalysis and neurobiology do not speak about the same thing. What links them is that both consider that human behavior partly relies on processes that escape consciousness. Nicolas Georgieff (professor of Child and Adolescent Psychiatry at Lyon-I University and member of the Institute of Cognitive Sciences) said that psychoanalysis and neuroscience represented "two profoundly original intellectual processes, two descriptions of the same object - the mechanisms of the human psyche—but on two opposite scales, with two methods incapable of answering the other's questions because they do not shed light on the same properties." Thus, psychoanalysis and neuroscience, far from being incompatible, can offer two levels of reading of a same mental illness, which can be considered as the result of a process of repression that must be brought to light and also as a biological disorder and therefore a neuronal failure.

In addition, for Francois Ansermet, psychoanalyst and head of a department of child psychiatry, and (Ansermet and Magistretti 2008), director of a psychiatric neuroscience center, it is possible to bring the two disciplines closer. They rely on the concept of brain plasticity that allows neuroscience to avoid burying itself in a too simplistic reductionist position only aimed at discarding any form of dualism, and at the same time allows the psychoanalytic and cognitive unconscious to coexist in the behavioral explanation of a subject.

Eve Suzanne said, Suzanne (2009) "In short, we can admit that we are not determined by our neurons, but we do shape them in the image of our life, although a part of our brain activity is determined by invariant processes."

8.8.2 Perspectives, by Eve Suzanne

Eve Suzanne also wrote: "Paul Ricoeur perfectly summarizes the problem we are facing with neuroscience, i.e., "My initial theory is that the discourses held on one side or the other belong to two heterogeneous perspectives, that is, one perspective cannot be fit into the other or derived from the other. One discourse addresses neurons, neuronal connections, the neuronal system, while the other is about knowledge, action, feeling, that is to say acts or states characterized by intentions, motives, values". Although Ricoeur is here at the level of semantic dualism, the same issue is also true as regards dualism of substances. Neurosciences consider that they managed to bridge the gap between the biological fact (neurons...) and the psychic fact (feeling...). To consolidate their claim, they need this presupposition since neurobiology aims to explain the very conditions of the emergence of the scientific idea. Science used to be only concerned about producing theories, with knowledge depending on the field in which it was practiced; it was not concerned about what it was itself produced from, out of which scientific theories were issued.

Thanks to synaptic plasticity, we can foresee the possibility for neuroscience to go beyond their aversion to any form of dualism and to mitigate their reductionist opinion. The functioning of the brain indeed simultaneously proceeds from brain processes identical for all and variations in our synapses in direct connection with our environment.

"The psychic Man and the neuronal Man, far from being enemies, are both indispensable to think of man as a whole."

References

F. Ansermet, P. Magistretti, *Freud au crible des neurosciences. Freud au crible des neurosciences*, pp. 84–85 (2008)

P. Buser, *Cerveau de soi (Cerveau de l'autre, Odile Jacob Paris)* (1998)

J.P. Changeux, P. Ricoeur, *Ce qui nous fait penser. La nature et la règle* (Editions Odile Jacob, Paris 1998)

M. Corcos, La mémoire et l'oubli, de la psychanalyse aux neurosciences. Carnet PSY **125**, 32–35 (2008)

S. Freud, *Remémoration, répétition et perlaboration* (La technique psychanalytique, PUF, 1994)

J. Godfrind, Transfert, compulsion et expérience correctrice. Revue française de psychanalyse **58**, 2 (1994)

A. Green, Préalables à une discussion sur la fonction de la théorie dans la formation psychanalytique. Revue française de psychanalyse **56**(2), (1992)

L. Naccache, *Le nouvel inconscient* (Odile Jacob, 2006)

J. Perkins, R.O. Sponholz, Stress- induced martensitic transformation cycling and two-way shape memory training on cu-zn-al alloys. Metall Trans A **15A**, 313–321 (1984)

P. Ricoeur, *La mémoire, l'histoire, l'oubli* (Editions du Seuil, Points Seuil, 2000)

E. Suzanne, *La psychanalyse face aux neurosciences. Implications philosophiques* (2009)

D.W. Winnicott, *La crainte de l'effondrement et autres situations cliniques (1989)* (Gallimard, 2000)

D.W. Winnicott, *Les aspects métapsychologiques et cliniques de la régression au sein de la situation analytique in "de la pédiatrie à la psychanalyse" Paris*. Payot pp. 331–348 (1954)

Chapter 9
Forgiveness

Abstract Denis Vasse (psychoanalyst) explains forgiveness. His purpose is in agreement with Paul Ricoeur one.

FORGIVENESS IS NOT FORGETFULNESS

People often say: "Come on, forget about it, turn the page, you will soon no longer remember it." Psychotherapy professionals know very well that we never forget. The so-called "forgotten" wounds have been buried in the unconscious and they keep torturing people (see the effects of non-forgiveness). We have to make them emerge again to be able to treat them. Forgiving does not mean forgetfulness, it means healing. We will be able to remember the event but we will no longer feel internal resentment. A scar does not hurt anymore. This is what happens when we forgive: we do not suffer anymore.

FORGIVING DOES NOT MEAN EXCUSING

Excusing means that we do not hold the offender accountable for his actions. We tend to find him extenuating circumstances. One explains his gesture or his words by the knowledge of his life. We minimize his actions or his words. In short, we protect him and we deny the harm done to us. But a fault is not excusable, even if it can be explained. A fault requires forgiveness. Did you know that when God presented himself to Moses, he presented himself as the merciful, the forgiving, but who does not hold the guilty for innocent?

FORGIVENESS IS NOT SYNONYMOUS WITH RECONCILIATION

Another misconception! What is it that establishes and maintains a relationship? Mutual trust. If trust is betrayed, it cannot return through mere will. Trust is earned, deserved, built, in this case has to be rebuilt. Two friends who severely hurt each other cannot decide that everything will go on as before at a finger snap. Reconciliation and forgiveness are not the same. Reconciliation comes in the wake of forgiveness, it is to be wished, but not systematically. As a result of an injury, we must decide whether we wish to carry on with the relationship, intensify it. Otherwise, it will simply stop because it has been broken.

© Springer Nature Switzerland AG 2019   73
C. Lexcellent, *Human Memory and Material Memory*,
https://doi.org/10.1007/978-3-319-99543-4_9

FORGIVENESS CANNOT BE IMPOSED

Forgiveness is an act of love ("for give"). The forgiving person must remain free of his/her choice. Forcing someone to forgive us comes down to telling him/her, "I want you to love me despite the nasty tricks I played on you". We can wish it and ask for it. We cannot compel the other to do it. Otherwise it is no longer forgiveness.

FORGIVENESS DOES NOT IMPLY GIVING UP ONE'S RIGHTS

Forgiveness is not meant to eliminate justice. A forgiven thief is not excused from returning stolen goods to their owners. Forgiveness does not remove the consequences of an unhappy act or word. Forgiving a murderer does not bring the victim back to life. Forgiveness is not an act of justice. It is a love process meant to rehabilitate the offender, his being. It dissociates him from the evil that inhabits him, and which led him to wrongdoing, so as not to condemn him with it.

FORGIVENESS DOES NOT CHANGE THE OTHER PERSON

When we forgive, something extraordinary happens that heals us and frees us. But we must not forgive thinking that this is what is going to change the other person. The offender may become aware of his behavior and change it and his inner attitude. But we have no power over the other person, who remains free and responsible to admit his fault or not.

This sentence seems quite debatable to me. Can there be forgiveness between the forgiver and the forgiven in the absence of a genuine encounter?

9.1   Denis Vasse (Psychoanalyst) and Forgiveness

According to the psychoanalyst Vasse (2000), it is difficult to speak of forgiveness. He means that it cannot be deduced from any logical reasoning. It has no reason.

If forgiveness obeys love, it never implies the abolition of law and justice (cf. the eternal "it does not matter"). On the contrary, if the arms that open to welcome us once again on the path of law and justice are those of Love, it is for us to be sent back to the school of the truth of life, where we were once carried away by the meanders of lies and death (Vasse 2007).

Forgiveness then appears as the "common horizon of accomplishment" of memory, history and forgetfulness, but it is by no means a "happy end". It can only be a matter of "difficult forgiveness", "neither easy nor impossible". It is more like a vow, an ideal towards which to tend. The author emphasizes the political function of a soothed memory and of forgiveness, wondering if politics does not begin where revenge ends, insofar as it would be counterproductive for a society to remain indefinitely angry against itself. The author wishes to warn his contemporaries by recalling at the very end of his work that it is only through mourning, guided by the horizon of reconciliation with the past and by the ideal of forgiveness that a society can definitively part from the past to make room for the future.

Then, as Jacques Lacan did, Denis Vasse delivers some meaningful formulas (Vasse, 2000).

"The fight against injustice has never made anyone just. What makes someone just is to give life."

As for the social aspect of the claim, asking is an act that implies trust.

In his talk on "forgiveness and original justice" Denis Vasse said, "forgiveness is in the right place when it reveals itself, it occupies the place of the signifier of the origin. It renews, it allows to speak, it gives life back to the one who forgives and to the one who is forgiven."

Forgiveness is the trace of a gift that comes neither from the executioner nor from the victim.

It is difficult to speak of forgiveness, it reveals the authentic educational skills of the law, the convicted person is referred to life that gives itself. Justice is the guarantee that the living being has the right to live.

The truth of our identity is not to have done foolish things, it is to be forgiven.

Forgiveness takes us away from the mortal process of repetition: the family story, phantasmagorical reality are no more than the reality of life. As if healing equaled to never having been ill!

Forgiveness is not a stuff to be used in any kind of circumstances. Yes, there are indeed unforgivable things, only reason calls on forgiveness.

Forgiveness is the justice of the living.

Forgiveness does not have to be justified. Forgiveness pierces through all imaginary images, touched at heart by words that give life back.

Forgiveness cannot be used in psychotherapy.

In no case is forgiveness economical or instrumental, it does not seek a strategy.

Those who are concerned about forgiveness, who want to reconcile you, are insufferable bores.

Loving oneself as "one among others" equals to loving oneself with others.

Forgiveness escapes the forgiver and the forgiven, it is a third word.

9.2 Paul Ricoeur and Forgiveness as a Beyond of the Deed

In his book "Memory, History, Forgetfulness" (Ricoeur 2000) Ricoeur developed the theme of forgiveness as an epilog entitled: "Difficult forgiveness". At the end of the book, Ricoeur tells us, "Beneath history lie memory and forgetfulness". Beneath memory and forgetfulness lies life, but writing life is another story: incompletion.

One can read his treatise on forgiveness in the light of the theme of life and recognition, but even more so in the light of a soothed memory and confidence in man, despite the fault. The theme of forgiveness refers to the theme of the "capable man"'s power to act and the aim of happiness.

Yet this recognition by Ricoeur was not naive optimism. Ricoeur made it clear that "forgiveness, if it makes sense, if it exists", makes up the common horizon of memory, history, and forgetfulness. Why "if it makes sense?" Because, he noted, forgiveness is difficult not only to give, to receive, but also to conceive. Ricoeur thus showed us right away where he stood in relation to forgiveness. Forgiveness is difficult. It is possible, but it is sometimes also "impossible" or "(almost) impossible". Let us note that this is quite in line with what Denis Vasse said earlier.

The author obviously positioned himself in the face of contemporary concerns about the meaning of forgiveness, notably those of Derrida (1999). He echoed the lights or shadows that the study of forgiveness may be under when one refers to the judicial and political spheres.

The question we are now going to develop concerns the fault made when forgiveness is described as a release. The purpose of the article (Fiasse 2007) was to show how to understand forgiveness as a beyond of action in Ricoeurian philosophy. Our reasoning will initially focus on the very notion of release. The philosophical stake will focus on the link between agent and act. We will then refer to the theme of gift economy. This second argumentative key will answer a utilitarian criticism of forgiveness. We will discuss the different issues of "gift economy" and we will shed light on how this theory allows us to take into account the need for confession in the forgiving phase, without reducing forgiveness to a model based on exchange. In the light of this analysis, we will see how Ricoeur took into account the objections raised by Jacques Derrida's impossible forgiveness (Derrida 1999). Lastly, this article will expose the practical dimension of the beyond of action in the judicial field, through the consideration due to the culprit, the theme of forgiveness gestures, and restorative justice.

9.2.1 Forgiveness as "A Power to Relieve the Agent from His Deed"

The term "forgiveness" refers to two different realities because we find ourselves in front of a pair: on the one hand, the person who lays down the act of forgiving a negative action that upset him/her, and on the other hand, the person who caused the injury and is likely to be forgiven and to word the request for forgiveness. Forgiveness, far from meaning forgetfulness of the fault, finds instead its starting point in the primordial recognition of the fault. Ricoeur is obviously aware of the excesses of a tendency toward victimization, and of the difficulties related to the recognition of the fault, if only because it is difficult for the people who admit their bad deed to word their wrongdoing.

He focused his interest on accountability, "this ability under which actions can be put to someone's account" (Ricoeur 2000, p. 596). "Accountability [...] this place where the agent binds to his deed and recognizes it as accountable" Ricoeur (2000, p. 594). But Ricoeur went further, since he associated accountability to responsibility. He declared, "There can indeed be forgiveness only where one can accuse someone, presume or declare them guilty."

Ricoeur noted that experiencing the fault affected the agent's action power. "Self-recognition is both action and passion, i.e., wrongdoing (action) and being affected by one's own action (passion)" (Ricoeur 2000, p. 598).

Ricoeur, after Klaus Kodalle, went so far as to quote the objection of the philosopher Hartmann (1926). Forgiveness would be a "moral fault" if it consisted in removing the guilty person from the bad deed.

Analyzing deeds, Hannah Arendt (Arendt 1983) emphasized in "La Condition de l'" "homme moderne" that the faculty to forgive is parallel to the faculty to promise.

Ricoeur also suggested another "decoupling" within the agent's power of action, i.e., between ability and its effectuation. This notion of capacity is indeed the second angle on which it is necessary to insist. Forgiveness is equivalent to believing in others, despite their actions. The disproportion must be maintained. There is a distance between the depth of the fault on one side and the height of forgiveness on the other.

Nevertheless, Ricoeur wanted to be able to take into account Jacques Derrida's objection, according to which a dissociation between the culprit and his act would mean that we forgive the guilty person while condemning their deed and thus that we forgive to an "other subject" than the one who committed the deed. "It is no longer the culprit as such that we forgive", according to Derrida (1999).

Like Hannah Arendt, who had not hesitated to turn to the New Testament to emphasize the experiences of forgiveness in the public sphere, Ricoeur also turned to Judeo-Christianity which praises forgiveness. While Hannah Arendt insisted on the human power to forgive, Ricoeur stressed its greatness, without expressly giving his opinion on the grace that leads to it. "There is forgiveness", the voice said (Ricoeur 2000, p. 594)".

The theme of the release of the agent of the deed, favored by Hannah Arendt's analysis, the non-reduction of human capacities to achievements, and the insistence on disposition for the good led Ricoeur to maintain a disproportion between guilt, the depth of the fault, and the height of forgiveness. As we are going to see, his analysis of forgiveness as a beyond of the deed should also be understood in the light of his ethics of gift and reciprocity.

9.2.2   Reciprocity and Gift Economy

Not only did Ricoeur aim to counter a reification of the agent to his evil deed, but he also wanted to place his ethics in an ethics of gift and reciprocity. We will show that he managed to avoid a utilitarian conception of the act in which one would only give in order to receive. He managed to take into account the unconditionality of forgiveness, extolled by Jacques Derrida, without ignoring the meaning of the request for forgiveness, through confession, repentance, even repair. The challenge was, therefore, to avoid reducing the request for forgiveness and the granting of forgiveness to a mercantile exchange, to a utilitarian profit, and at the same time to avoid neglecting the realistic aspect of forgiveness. It is indeed easier to forgive the one who recognizes his fault and it also belongs to the dignity of human beings to be expected to admit their wrongdoing.

Ricoeur, therefore, wanted to hear criticisms of only conditional forgiveness, which is why he intended to maintain the unconditionality of forgiveness, which he also called the height of forgiveness. But at the same time, he emphasized the value of considering forgiveness as in a bipolar relationship. In other words, yes, it makes sense to take into account confession, repentance, and even repair. There exists a certain equation of confession and forgiveness that cannot be reduced to exchange.

Faced with the criticism of a utilitarian relationship, Ricoeur maintained that the confession that leads to forgiveness should not be conceived according to a model of exchange in the same way as in a system of obligation, i.e., to give, to receive, and to return. This is why he considered that asking for forgiveness' " is also being ready to receive a negative answer: no, I cannot, I cannot forgive" (Ricoeur 2000, p. 626).

9.2.3   The Incognitos of Forgiveness at the Judicial Level, Restorative Justice

It is not uncommon to see the theme of justice as an obstacle to the theme of forgiveness. Although this point has already been dealt with about the release of the deed, it is necessary to further investigate the relationship between forgiveness and justice considering the extent of the debates related to this issue. The questions of the public opinion, when the theme of forgiveness comes into play, are the following: would forgiving mean forgetfulness? Erasing injustice? If forgiveness is possible, does it lead to the belief that the perpetrator of crimes should not be judged?

It is, therefore, necessary to distinguish between the rules of justice—which also encompass the common good—on the one hand, and the depths of one's soul on the other hand. As Paul Ricoeur put it, people forgive people, courts do not forgive. The fault turns out to be unforgivable by right. The breach of common rules leads to the punishable. If forgiveness consists in ruling out punishment, then it is impossible within the institutional framework because it would lead to lifting the punitive sanction, in other words to "not punishing where and when we can and must punish". Forgiveness would create impunity, which would be a great injustice. This is why Ricoeur took care to distinguish forgiveness from amnesty. Amnesty belongs to the political body. It appears as an antithesis of forgiveness, since it is forgetfulness, while forgiveness requires memory. However, Ricoeur did not fail either to speak about the incognitos of forgiveness, in Klaus Kodalle's words, in other words, a spirit of forgiveness at the judicial level.

The articulation of justice and the spirit of forgiveness can be found in the three poles of the judiciary, i.e. at the levels of the law, the victim, and the convicted person. In his article "The just, justice and its failure", Ricoeur expressed his major concern to see the limits of the violence exerted by justice on behalf of the State. He insisted on the importance of "saying a word of justice in the name of the people", but he problematized the idea of inflicting suffering in the name of the State, which he also called the strenuousness of the sentence.

The spirit of forgiveness in justice, therefore, consists in eradicating the sacred component of vengeance. As regards the right to security, Ricoeur invited his readers to think of it not only as a potential victim, "but perhaps also as a potential offender."

He ended his study by expressing a "dream" of a final appearance of the offended together with the offender. This is completely consistent with what Denis Vasse wrote in "Vengeance et Pardon" (Vasse 2000).

This dream is obviously linked to a justice called restorative justice, or also rehabilitating, reconstructive justice. In "Memory, History, Forgetfulness", Ricoeur mentioned the atypical case of the "Truth and Reconciliation" Commission in South Africa, which he considered as an example of restorative justice.

9.2.4 Conclusion of This Section

In conclusion, we saw that Ricoeur did not want to underestimate wrongdoing (Fiasse 2007). The depth of the fault, which is unforgivable by right, is maintained in the face of the height of forgiveness. Ricoeur spoke of forgiveness as a release of the deed. Although guilt is not denied, the perpetrator is not reduced to the negative action he committed. The words of forgiveness, when pronounced, state that the agent is better than his deeds. He has returned to his ability to innovate, as shown by Arendt (1983). In answer to Derrida (1999), who saw in the repentant a better person than the author of the crime, Ricoeur insisted on the different dimensions of the capable man; he considered that the power to act was not limited to its different effectuations, even if he admitted the radicalism of guilt. Forgiveness, therefore, calls upon the resources of regeneration of the self and the background of goodness present in man.

This beyond of the deed is also favored by gift economy in Ricoeurian ethics, even if it remains in an optative mode. Ricoeur used the golden rule and extreme commandment to "love one's enemies" to inscribe forgiveness in an ethics of overabundance. The asymmetry established by the golden rule between the agent and the patient is found in the contrast between forgiveness and fault. The commandment to love one's enemies, an extreme version of the golden rule, establishes a point of rupture with a utilitarian vision of acting. It also makes it possible to take into account the unconditionality of forgiveness, underlined by Jacques Derrida, but it does not ruin the hope of a reciprocal relationship. In the same way, the confession that leads to forgiveness cannot be conceived as a trading, but the hope that it may be pronounced and received remains.

From a judicial point of view, the beyond of the deed must not lead to impunity, but promote consideration for the culprit and renouncement to revenge. It can only be conveyed by gestures. The case of restorative justice testifies to a possible exchange between victims and inmates in which the fault is maintained, the necessity of confession is admitted, and soothing is conceived and hoped for as another incognito of the spirit of forgiveness.

References

H. Arendt, *Condition de l'homme moderne (trad. G. Fradier)* (Calmann-Lévy, 1983)

J. Derrida, Le siécle et le pardon. *Le Monde des Débats* 12 (1999)

G. Fiasse, Paul Ricoeur et le pardon comme au-delà de l'action. Laval thèologique et philosophique **63**(2), 363–376 (2007)

N. Hartman, *Ethics t.III Moral Freedom* (Macmillan, New York, 1926)

P. Ricoeur, *La mémoire, l'histoire, l'oubli* (Editions du Seuil, Points Seuil, 2000)

D. Vasse, Vengeance et pardon ens lyon (2000)

D. Vasse, Le pardon et le sentiment de culpabliité. Christus **216**, 464–470 (2007)

Chapter 10
Post-memory

Abstract The "post-memory" is defined as this underground and enigmatic memory, both intimate and collective which characterizes the transmission of historical trauma to generations that did not leave it.

In the very recent issue of October 2017 of the journal "Esprit" ("Spirit"), Marianne Hirsch, Veronica Estay Stange, and Nathalie Bittinger investigated the forms of post-memory. In the introduction to the theme "Haunted by Memory," Jonathan Chalier defined post-memory as this underground and enigmatic memory, both intimate and collective, which characterizes the transmission of historical trauma to generations that did not live it. What are the secret routes it uses to perpetuate itself and maintain its effects among the descendants of those who suffered it? (Chalier 2017). Marianne Hirsch's concept of post-memory refers to the relationships that generations have with traumatic experiences that they did not experience directly. As she writes, "Events happened in the past, but their effects are still felt in the present" (Hirsch 2017). "The recent development of epigenetic science seems to provide an empirical approach to phenomena that thwart common representation. Chalier (2017)"Epigenetics is the study of cellular or phenotypic variations caused by external environmental factors, which do not affect the DNA sequence itself, but the expression of the gene. It is now considered that epigenetic marks caused by environmental factors are not lost across generations, but are inherited and transmitted with the same DNA sequences" (Hirsch 2017).

At this point, I cannot help but think of Patrick Modiano's first novels. Focusing on interiority, repetition and nuance, Patrick Modiano's novelistic work approaches a form of autofiction through his quest for lost youth. It focuses primarily on Paris under the Occupation (while he was born on July 30, 1945!). It focuses on depicting the lives of ordinary individuals confronted with the tragedy of history and acting in a random or opaque way.

In the post-memory inquiry, we must all think about our vulnerability to the violence of history together, and about our ability to repair through the stories we tell ourselves, to find ways of expression that can heal these secret wounds.

In her own work on post-memory, Mariane Hirsch argued that the intergenerational transmission of trauma went beyond individual and family limits, and that it depended on acts of transmission that are embodied, emotional and symbolic, and also complex and numerous (Hirsch 2017).

For example, Yehuda et al. (1998) found that the epigenetic marks of children of Holocaust survivors showed patterns of stress hormones that predisposed them more to post-traumatic stress disorder and other disorders than their peers in the control group.

In her article, ≪Survivre à la survie; remarques sur la post-mémoire≫ (Estay 2017) Veronica Estay Stange stated, "I have recollections that do not belong to me. I carry the nostalgia of others." This is how she described, a long time ago, her relationship to Chilean history since the coup d'etat that took place in 1973, which she did not live since she was born much later.

According to Hirsch (2017), "the attention devoted to epigenetics is a recent development; it forestalls the study of individual and collective memory and post-memory that corroborates the testimony of those who inherited the traumas of previous generations. But epigenetics, as a generalized paradigm, can also be the source of difficulties. In popular imagination, at least, the results promised by genetic and epigenetic research are endowed with a force of truth that is denied to other paradigms."

Indeed, this "proxy memory" characteristic of survivors' descendants is also an "infra-memory" that questions traditionally recognized mechanisms, especially the narrative and the other the past (Estay 2017). Paul Ricoeur's identification of two dimensions of remembrance, i.e., cognitive and pragmatic, is enlightening in this respect, "To remember is to have a recollection or to start a quest for a recollection" (Ricoeur 2000).

As for survival, "You are ashamed, Primo Levi wonders, because you are alive instead of another person?" (Lévi 1989).

"As opposed to infra-memory, post-memory can be defined as the actual, reconstructed and expressed utterance in the third person of an inherited absent or forbidden enunciation" (Estay 2017).

In the context of the 1973 coup d'état in Chile, it is not a question of bringing everything back to presence by showing horror as such, but of pointing out the place of an ultimate irrepresentable because of its excess" (Estay 2017).

Nathalie Bittinger made the classic observation that the world wars, the Holocaust, genocides (in Rwanda, Cambodia, ...), twentieth century totalitarianism have engendered a traumatic memory. Therefore, how is it possible to fight against tragedy and resentment, aphasia or logorrhea, the repression or untimely memory carried by these moments of barbarity (Bittinger 2017)? One can oscillate between verbal delirium and mutism. Only narrative and formalization offer their resources to heal a traumatic memory that does not stop. "In the aftermath of the event, one of the most forceful exorcisms is thus offered by representation, provided that it is fair, honest and anti-spectacular" (Bittinger 2017).

Stéphane Audoin-Rouzeau's latest book, entitled "Une Initiation", takes a step aside on the memory denial that cost the lives of at least 800,000 Tutsis in Rwanda between April and July 1994. He underlines that "in France, we are still in a denial of the French responsibility for this genocide" (Audoin-Rouzeau 2017).

What of the question "never again"? Since 1988–1989, trials have been following one another, and all the survivors of the Holocaust testified on television and radio shows so that the same would not happen again. Yet, the genocide of the Rwandan Tsutsis in the XXth century seems to be the only mass massacre that could have been prevented, while we live in a society that claims to prevent any repetition of a genocide like that of the European Jews (Audoin-Rouzeau 2017).

In short, what have we done with the "duty of memory"?

References

S. Audoin-Rouzeau, Hantés par la mémoire: Balayer les cendres. ESPRIT **438**, 81–96 (2017)

N. Bittinger, Hantés par la mémoire; Chanter les martyrs en Chine et à Taiwan. ESPRIT **438**, 73–80 (2017)

J. Chalier, Hantés par la mémoire: Introduction. ESPRIT **438**, 40–41 (2017)

V. Estay, Stange. Hantés par la mémoire: survive à la survie, remarques sur la post-mémoire. ESPRIT **438**, 62–72 (2017)

M. Hirsch, Hantés par la mémoire: Ce qui touche à la mémoire. ESPRIT **438**, 42–61 (2017)

P. Lévi. *Les Naufragés et les Rescapés. Quarante ans après Auschwitz., traduit par André Maugé* (Gallimard, Paris, 1989)

P. Ricoeur, *La mémoire, l'histoire, l'oubli* (Editions du Seuil, Points Seuil, 2000)

R. Yehuda, J. Schmeidler, M. Wainberg, K. Binder-Brynes, T. Duvdevani, Vulnerability in post-traaumatic stress disorder in adult offspring of holocaust survivors. Am. J. Psychiatry **155**(9), 1163–1171 (1998)

Chapter 11
Conclusion and General Summary

The present text is like a journey through memory. My knowledge of the memory of certain solid materials led me to seek a continuity with human memory. Let us say it plainly, the memory of materials is easier to understand, since creating defects in their microstructure is sufficient to make them acquire memory.

These defects may be assimilated to memory traces in the human brain.

However, the mechanisms of acquisition of human memory appear to be rather complex, as shown by neuroscience.

Let us say that the mechanisms associated to the training of shape memory alloys are more basic than those occurring in the brain.

The knowledge of human memory dates back to at least the Greek civilization, if not before. The scientific knowledge of shape memory alloys goes back to Martens, a German scientist, around 1950. He gave his name to the "martensitic transformation", which is the key to SMA behavior. The common point between materials and humans is the crystalline defects generated in SMAs during their training and the memory traces in the human brain carrying the recollection.

The big difference between these materials and humans is that materials have no subconscious, whereas we may well think that humans have one!

In short, have I established a continuity between human memory and the memory of materials? Some analogies have been emphasized between defects and traces, for example.

This question remains open today. In my opinion, there is a distance between neuroscience, which is often a technical approach to memory, and "human memory".

Let us say that human memory and memory in the physical or mathematical sense are not of the same nature. There is no neutral memory!

Thus, I sketched a brief history of memory from the ancient Greeks to the present day. Phenomenological, philosophical, and technical approaches (see neuroscience) have been described.

© Springer Nature Switzerland AG 2019
C. Lexcellent, *Human Memory and Material Memory*,
https://doi.org/10.1007/978-3-319-99543-4_11

A decisive contribution was brought by Paul Ricoeur's famous work "Memory, History, Forgetfulness" published in 2000. Given the energy he devoted to forgiveness, I hardly understand why he did not add this theme to its title. His words about forgiveness appear to me in symbiosis with those of the psychoanalyst Denis Vasse.

The pinnacle of the humanity of memory (or of a happy or soothed memory) lies in forgiveness. Of course forgiveness has nothing to do with SMAs!

Paul Ricoeur made a rather sharp criticism of the neuroscience approach.

Apart from his exchanges with the biologist Jean Pierre Changeux, I found a few comments from neuroscientists about Paul Ricoeur's work (Ricoeur 2000). Francis Eustache and his collaborators, although their book "Ma mémoire et les autres" (Eustache et al. 2017) is devoted to the entanglement between individual memory and collective memory, did not make any reference to the writings of Ricoeur (2000) on this subject.

In recent research, we find exotic questions such as "does the thought of death contribute to memory being encoded with a survival scenario?" (Bugaiska et al. 2014) or "what if musical skill, talent, and creativity were a matter of memory organization and strategies?" (Drai-Zerbib 2016).

Right now, I leave these questions open to the reader's sagacity.

Finally, I just discovered the concept of post-memory and memristor.

One last open question: animals, trees, etc., appear to have memory, like metal materials. What does not have memory?

References

P. Ricoeur, *La mémoire, l'histoire, l'oubli*. Editions du Seuil, Points Seuil (2000)

F. Eustache, H. Amieva, C. Thomas-Anterion, J.G. Ganascia, R. Jaffard, C. Peschanski, B. Stiagler, *Ma mémoire et les autres*. ESSAI LE POMMIER! (2017)

A. Bugaiska, M. Mermillod, P. Bonin, Does the thought of death contribute to the memory benfit of encoding with a survival scenario? *Memory* (2014)

V. Drai-Zerbib. What if musical skill, talent and creativity were just a matter of memory organisation and strategies? Int. J. Talent Dev. Creat. (**4**(1), August, 2016, **4**(2) December, 2016), 87–95 (2016)

Printed in the United States
By Bookmasters